Energy Policy Act of 2005

Summary and Analysis of the Act's Major Provisions

Kevin J. McIntyre
Martin V. Kirkwood
Jason F. Leif

General Editors

Energy Delivery and Power Practice
Oil and Gas Practice
Jones Day

For further information on energy legislation and regulation, see *Energy Law and Transactions* (David J. Muchow & William A. Mogel, eds., LexisNexis Matthew Bender)

LexisNexis™
Matthew Bender®

QUESTIONS ABOUT THIS PUBLICATION?

For questions about the **Editorial Content** appearing in these volumes or reprint permission, please call:

Linda J. Folkman, J.D. at .... 1-800-252-9257 (Ext. 2249)
Internet Address: .... Linda.Folkman@lexisnexis.com
Patrick E. Cannon, J.D. at .... 1-800-424-0651 (Ext. 3238)
Internet Address: .... Patrick.Cannon@lexisnexis.com
Outside the United States and Canada please call .... (973) 820-2000

For assistance with replacement pages, shipments, billing or other customer service matters, please call:

Customer Services Department at .... (800) 833-9844
Outside the United States and Canada, please call .... (518) 487-3000
Fax number .... (518) 487-3584
Customer Service Website .... http://www.lexisnexis.com/custserv/

For information on other Matthew Bender Publications, please call
Your account manager or .... (800) 223-1940
Outside the United States and Canada, please call .... (518) 487-3000

ISBN 0-8205-6985-2

LIBRARY OF CONGRESS CATALOGING IN PUBLICATION DATA

Energy Policy Act of 2005 : summary and analysis of act's major provisions / Kevin J. McIntyre, Martin V. Kirkwood, Jason F. Leif, general editors.
p. cm.
Includes index.
ISBN 0–8205–6985–2 (softbound)
1. Energy industries—Law and legislation—United States. 2. Energy conservation—Law and legislation—United States. 3. Energy policy—United States. I. McIntyre, Kevin J. II. Kirkwood, Martin V. III. Leif, Jason F. IV. Title.
KF2120.E543 2006
346.7304'679--dc22

2006001911

Editorial Offices
744 Broad Street, Newark, NJ 07102 (973) 820-2000
201 Mission St., San Francisco, CA 94105-1831 (415) 908-3200
www.lexis.com

MATTHEW BENDER

(Rel.30A–2/06 Pub.336)

Statement on Fair Use

LexisNexis Matthew Bender recognizes the balance that must be achieved between the operation of the fair use doctrine, whose basis is to avoid the rigid application of the copyright statute, and the protection of the creative rights and economic interests of authors, publishers and other copyright holders.

We are also aware of the countervailing forces that exist between the ever greater technological advances for making both print and electronic copies and the reduction in the value of copyrighted works that must result from a consistent and pervasive reliance on these new copying technologies. It is LexisNexis Matthew Bender's position that if the ''progress of science and useful arts'' is promoted by granting copyright protection to authors, such progress may well be impeded if copyright protection is diminished in the name of fair use. (See *Nimmer on Copyright* §13.05[E][1].) This holds true whether the parameters of the fair use doctrine are considered in either the print or the electronic environment as it is the integrity of the copyright that is at issue, not the media under which the protected work may become available. Therefore, the fair use guidelines we propose apply equally to our print and electronic information, and apply, within §§107 and 108 of the Copyright Act, regardless of the professional status of the user.

Our draft guidelines would allow for the copying of limited materials, which would include synopses and tables of contents, primary source and government materials that may have a minimal amount of editorial enhancements, individual forms to aid in the drafting of applications and pleadings, and miscellaneous pages from any of our newsletters, treatises and practice guides. This copying would be permitted provided it is performed for internal use and solely for the purpose of facilitating individual research or for creating documents produced in the course of the user's professional practice, and the original from which the copy is made has been purchased or licensed as part of the user's existing in-house collection.

LexisNexis Matthew Bender fully supports educational awareness programs designed to increase the public's recognition of its fair use rights. We also support the operation of collective licensing organizations with regard to our print and electronic information.

Energy Policy Act of 2005

Summary and Analysis of the Act's Major Provisions

Kevin J. McIntyre
Martin V. Kirkwood
Jason F. Leif

General Editors

Energy Delivery and Power Practice
Oil and Gas Practice
Jones Day

* * *

	Contributors
Title I—Energy Efficiency (§ 1.01)	Sameh I. Mobarek
Title II—Renewable Energy (§ 1.02)	Kenneth B. Driver Brian J. McManus
Title III—Oil and Gas (§ 1.03)	Jason F. Leif Joshua L. Fuchs Daniel Lim Elizabeth A. Jackson
Title IV—Coal (§ 1.04)	Kevin J. McIntyre
Title V—Indian Energy (§ 1.05)	Shelby L. Provencher
Title VI—Nuclear (§ 1.06)	Kevin J. McIntyre
Title VII—Vehicles and Fuels (§ 1.07)	Sameh I. Mobarek

Contributing Authors (continued)

	Contributors
Title VIII—Hydrogen (§ 1.08)	Kristiana A. Garcia
Title IX—Research and Development (§ 1.09)	Shelby L. Provencher
Title X—Department of Energy Management (§ 1.10)	Jonathan F. Christian
Title XI—Personnel and Training (§ 1.11)	Jonathan F. Christian
Title XII—Electricity (§ 1.12)	Martin V. Kirkwood William J. Harmon
Title XIII—Energy Policy Tax Incentives (§ 1.13)	Todd Wallace Mark E. Patrick
Title XIV—Miscellaneous (§ 1.14)	Laurie M. Earl
Title XV—Ethanol and Motor Fuels (§ 1.15)	Laurie M. Earl
Title XVI—Climate Change (§ 1.16)	Kristiana A. Garcia
Title XVII—Incentives for Innovative Technologies (§ 1.17)	Brian J. McManus
Title XVIII—Studies (§ 1.18)	Peter I. Trombley

Energy Policy Act of 2005

Summary and Analysis of Act's Major Provisions

Scope

This chapter summarizes the Energy Policy Act of 2005 (Act), which was signed into law by President Bush on August 8, 2005.[1] The Act is the first piece of comprehensive energy legislation passed by Congress since the Energy Policy Act of 1992 (1992 Act).[2] Like the 1992 Act, its 2005 counterpart is far-reaching and will likely impact the development, use, and regulation of energy resources and the energy industry for years to come.

Several policy goals are apparent from the Act. First, through numerous provisions Congress sought to decrease America's dependence on foreign energy sources by encouraging domestic production of both conventional and alternative fuels. Consistent with this goal, Congress also adopted a series of measures designed to get consumers and businesses to conserve fuel. The new provisions also streamline or simplify the development or construction of energy projects to encourage investment in the nation's energy infrastructure and thereby ensure that energy resources get to market quicker. Second, Congress used the Act to continue to foster competition in energy markets. However, it is also clear that Congress learned a lesson from the competitive abuses of the late 1990s and early 2000s and, as a result, the Act gives federal regulators of the wholesale electric and gas markets new tools and penalty authority to deter and penalize manipulative

[1] Pub. L. No. 109-58, 119 Stat. 594 (2005).

[2] *See* Chapter 58 of Energy Law and Transactions (David J. Muchow & William A. Mogel eds., LexisNexis Matthew Bender), for a summary of the Energy Policy Act of 1992.

and other improper behavior. With these changes, Congress expects competitive, but fair, markets to continue to develop.

The Act consists of 18 titles. Below is a short summary of the major provisions that are explained in more detail in this chapter.

Oil and Gas

The Oil and Gas provisions are designed to foster the development of domestic oil and gas supplies. For example, the Act implements a variety of production incentives to increase production from marginal and difficult-to-access oil and gas resources. In addition, Congress sought to ensure adequate supplies of liquefied natural gas (LNG) by giving the Federal Energy Regulatory Commission (FERC) clear and primary jurisdiction over the permitting of LNG import terminals. Finally, in response to the competitive abuses in the energy industry over the past several years, Congress enhanced the FERC's penalty authority for violations of the Natural Gas Act (NGA) and the Natural Gas Policy Act (NGPA) and gave the FERC new authority to deter and punish those that seek to manipulate wholesale gas markets.

Coal

Title IV of the Act establishes a Clean Coal Power Initiative to develop new clean coal technology, with a particular emphasis on research and development of coal gasification technology. The Act also establishes a separate Clean Air Coal Program that provides funds both for state-of-the-art generation projects and air quality enhancement projects. It is anticipated that the funding for the Clean Coal Power Initiative and the Clean Air Coal Program will encourage the use of coal, America's most abundant native fossil fuel, in an environmentally sound manner, and thereby lessen the Nation's reliance on foreign resources. Domestic coal production incentives are provided as well, primarily through an expansion of leasing rights on federal lands.

Nuclear

The Act indicates Congress's clear desire to increase the Nation's reliance on nuclear energy as a means to lessen reliance on foreign oil. The Act calls for the creation and construction of advanced nuclear power generating facilities. For example, the Act provides an economic incentive for the construction of such nuclear facilities and certain protections to owners of the plants from delays or other actions that may interfere with the construction or operation of these facilities. The Act also takes steps to improve the security of the Nation's nuclear facilities and to address security concerns pertaining to nuclear fuel and other radioactive materials, including mandatory new Nuclear Regulatory Commission (NRC) security review procedures and stringent new controls over nuclear equipment and materials and other radiation sources.

Electricity

Congress repealed the Public Utility Holding Company Act of 1935 in an effort to encourage investment in the Nation's bulk power grid and to foster competition. However, the Act replaced the repealed 1935 Act with a new law that imposes much of the same reporting and recordkeeping requirements on holding companies. To further encourage competition, Congress terminated the requirement under the Public Utility Regulatory Policies Act of 1978 (PURPA) that required public utilities to purchase from, and sell to, cogeneration and small power production facilities if those entities have access to competitive markets.

To ensure the reliable operation of the Nation's bulk power grid, FERC must designate an Electric Reliability Organization that will be tasked with developing and implementing mandatory reliability standards. In addition, to foster investment in the high voltage transmission grid in areas facing congestion, the Secretary of Energy must perform studies on a regular basis to determine areas of congestion and FERC will have "backstop" permitting authority to ensure that transmission facilities will not be prevented by recalcitrant state or local governments.

Finally, in reaction to competitive abuses, Congress gave FERC additional tools and expanded penalty authority to deter and punish market participants that manipulate the market or engage in other improper behavior.

Alternative Fuels and Renewables

Several provisions of the Act seek to lessen the Nation's reliance on foreign resources by developing alternative and renewable fuels. The Act provides incentives to develop and use renewable energy resources, including geothermal, hydroelectric, and biomass resources. The Act encourages the use of alternative fuels in vehicles, development of hybrid vehicles, and automobile efficiency. The Act also contains separate provisions to promote the use of ethanol in motor fuels. Finally, the Act provides for research, development and the commercialized use of hydrogen as an alternative fuel source.

Energy Efficiency

The Act seeks to promote energy efficiency through creation of new federal programs, the production of energy efficient products, and through the use of renewable fuels and energy efficient products in public buildings.

SYNOPSIS

§ 1.01 Title I—Energy Efficiency

[1] Scope

Congress has recognized that increased energy efficiency could reduce demand and thereby lessen the need for new fuel sources and facilities. Therefore, in Title I, Congress established various programs designed to foster the efficient use of energy and increased conservation.[1]

[2] Federal Programs

[a] Energy and Water Savings Measures in Congressional Buildings

Energy efficiency of buildings can be improved through the use of different insulating materials in building design as well as other design concepts, such as solar orientation and shade tree landscaping. To institute some of these concepts in congressional buildings, Section 101(a) of the Act amends Part 3 of Title V of the National Energy Conservation Policy Act (NECPA) to require the Architect of the Capitol to develop and implement an energy and water conservation plan for congressional buildings. The plan should be consistent with requirements of other federal buildings. Congress appropriated $2 million for this endeavor, and required the Architect to submit to it an annual report on the plan's implementation.[2]

[b] Energy Management Requirements

In Section 102(a) of the Act, Congress amended the NECPA to change the baseline year for calculating the energy savings for federal buildings from 1985 to 2006. The amendment also established a target of 20% reduction in energy consumption per gross square foot to be achieved in 2% increments between 2006-2015. As an incentive to achieve this target, federal agencies can retain funds allocated for energy expenses that are not incurred because of energy and water conservation. However, the agencies can use these savings only for energy efficiency, water conservation, or unconventional and renewable energy resources projects. Congress also required the Secretary

---

[1] Congressional Research Service, CRS Issue Brief IB10143, *Energy Policy: Comprehensive Energy Legislation (H.R. 6) in the 109th Congress* at 11 (July 29, 2005).

[2] 2005 Energy Policy Act § 101(a).

of Energy to review the results of the requirements under this section in 2014 and to recommend new targets for the 2016-2025 period.[3]

[c] Energy Use Measurement and Accountability

Section 103(a) requires that all federal buildings have advanced meters or metering equipment installed by October 1, 2012, to provide daily and hourly data on the buildings' electricity consumption. Each federal agency is required to submit a plan, within six months of the date guidelines are established for installation of the meters or metering equipment, detailing the steps to take for compliance.[4]

[d] Procurement of Energy Efficient Products

The Act amends the NECPA to require federal agencies to incorporate into the specifications for all procurements involving energy products and systems criteria for energy efficiency that are consistent with the criteria used for rating Energy Star products and products rated by the Federal Energy Management Program (FEMP). Energy Star products are products rated under the Energy Star program, as established by Section 324A of NECPA, and FEMP-rated products are products rated to be among the highest 25% of equivalent products for energy efficiency. However, federal agencies are exempt from this requirement if: (1) the Energy Star or FEMP-rated product is not cost-effective over the life of the product taking the energy cost savings into account; or (2) no Energy Star or FEMP-rated product is reasonably available to meet the functional requirements of the agency.[5]

[e] Energy Savings Performance Contracts

The Act amends Section 801(c) of NECPA to give federal agencies until October 1, 2016, to enter into contracts to achieve energy savings and related benefits.[6] Furthermore, Congress ratified all contracts entered into for that purpose after October 1, 2003, which is the current sunset date for agency authority to enter into such contracts, and before the enactment of the 2005 Energy Policy Act.[7]

---

[3] 2005 Energy Policy Act § 102(a)–(b).

[4] 2005 Energy Policy Act § 103(a).

[5] 2005 Energy Policy Act § 104(a).

[6] 2005 Energy Policy Act § 105(a).

[7] 2005 Energy Policy Act § 105(b).

[f] Voluntary Commitments to Reduce Industrial Energy Intensity

Section 106 authorizes the Secretary of Energy to enter into voluntary agreements with industrial customers that consume significant amounts of energy for each unit of output to reduce their energy demand per unit of output by 2.5% annually for the 2007-2016 timeframe.[8] Section 106 further authorizes the Department of Energy (DOE) to provide technical assistance grants to all industrial customers electing to enter into such voluntary agreements to assist achieving this goal.[9] This authority formalized efforts by other agencies which had industrial energy efficiency programs without express congressional authority to initiate them, e.g., the now-expired Climate Wise Program of the Environmental Protection Agency (EPA).

[g] Advanced Building Efficiency Testbed

Section 107(a) requires the DOE, in conjunction with the General Services Administration (GSA), to establish a program for the development, testing, and demonstration of advanced engineering systems, components, and materials to enable innovations in building technology. The objective of the program is to evaluate and demonstrate efficiency concepts for government and industrial buildings.[10] The program must be led by a partnership of universities that have the expertise required to achieve the program's goals.[11] Congress appropriated $6 million for the 2006-2008 period to carry out this program.[12]

[h] Increased Use of Recovered Mineral Components in Federally-Funded Projects Involving Procurement of Cement or Concrete

The Act amends Subtitle F of the Solid Waste Disposal Act[13] to promote the increased use of recovered mineral components in federally-funded highway or other transportation facilities, or in a

---

8 2005 Energy Policy Act § 106(b).

9 2005 Energy Policy Act § 106(e).

10 2005 Energy Policy Act § 107(a).

11 2005 Energy Policy Act § 107(b).

12 2005 Energy Policy Act § 107(c).

13 42 U.S.C. § 6961 *et seq.*

federal, state, or local government building or other public facility that involves the use or procurement of cement or concrete.[14] The Act also requires a study and report on the extent to which the procurement practices result in energy savings and benefits.

[i] Federal Building Performance Standards

Section 109 of the Act amends the Energy Conservation and Production Act (ECPA),[15] to revise the standards developed for federal buildings to require economically-justified energy efficiency measures. In addition, Section 109 requires the DOE to change the standard against which energy efficiency gains are measured within one year from date of enactment of the Act.[16]

[j] Daylight Savings Time

In an effort to reduce energy used for night-time electric lighting, Section 110(a) extends daylight savings time by four weeks. Beginning in 2007, daylight savings time will begin on the second Sunday in March and end on the first Sunday in November, instead of beginning on the first Sunday in April and ending on the last Sunday in October.[17]

[k] Enhancing Energy Efficiency in Management of Federal Lands

As a policy matter, Congress has determined that federal agencies should enhance the use of energy efficient technologies in the management of natural resources. Towards that end, Section 111(b) requires the Departments of the Interior, Commerce, and Agriculture to incorporate energy efficient technologies in their public and administrative buildings.[18] The Departments must also use, to the extent practicable, energy efficient vehicles, including vehicles equipped with biodiesel or hybrid engine technologies.

---

14 2005 Energy Policy Act § 108(a).

15 Pub. L. No. 94-385 (1976), codified at 42 U.S.C. § 6801 *et seq.*

16 2005 Energy Policy Act § 109.

17 2005 Energy Policy Act § 110(a).

18 2005 Energy Policy Act § 111(b).

[3] Energy Assistance and State Programs

[a] Low Income Home Energy Assistance Program

Section 121 appropriates $5.1 billion for each of fiscal years 2005-2007 for the Department of Health and Human Services to provide grants to states to assist low income households in paying for energy costs. This appropriation represents a $3.1 billion increase in the amounts annually provided over those made through fiscal 2004. The Act encourages states to purchase renewable fuels, including biomass, to fulfill their obligations to receive the grants.[19]

[b] Weatherization Assistance

Consistent with Congress' intent to improve the energy efficiency of low income housing, Section 122(a) appropriates $500 million, $600 million, and $700 million for fiscal years 2006, 2007 and 2008, respectively, to provide grants to "weatherize" such dwellings.[20] Section 122(b) amends the ECPA to broaden the category of persons entitled to weatherization assistance by increasing the eligible income threshold from 125% of the poverty level to 150%.[21]

[c] State Energy Programs

The Energy Policy and Conservation Act (EPCA) requires the DOE to develop an energy conservation plan under which states will improve energy use by 10% or more in 2000 as compared to 1990. Section 123(b) of the 2005 Energy Policy Act amends EPCA to increase this target saving rate to 25%, which is to be achieved by 2012 as compared to the 1990 level.[22] In support of this objective, Congress appropriated $100 million for each of fiscal years 2006 and 2007, and $125 million for fiscal year 2008.[23]

[d] Energy Efficient Appliance Rebate Program

Congress created an incentive for states to develop energy efficient appliance programs that provide rebates to residential consumers for

[19] 2005 Energy Policy Act § 121(a)–(b).

[20] 2005 Energy Policy Act § 122(a).

[21] 2005 Energy Policy Act § 122(b).

[22] 2005 Energy Policy Act § 123(b).

[23] 2005 Energy Policy Act § 123(c).

the purchase of residential Energy Star products to replace used appliances of the same type. For states that develop and implement such programs, Section 124(f) of the Act appropriates $50 million for each of fiscal years 2006-2010 to be distributed in proportion to the population of each state as compared to the total population of all states that have such programs.[24] The funds are intended to supplement, not supplant, state funds supporting these programs.[25]

[e] Energy Efficient Public Buildings

The Act authorizes the DOE to provide grants to states to improve the energy efficiency of public buildings and facilities.[26] Section 125(c) appropriates $30 million for each of fiscal years 2006-2010 for that purpose.[27]

[f] Low Income Community Energy Efficiency Pilot Program

Section 126 of the Act authorizes the DOE to provide grants to units of local government, private, non-profit community development organizations, and Indian tribe economic development entities to identify and develop alternative, renewable, and distributed energy supplies and to support projects that increase energy conservation in low income rural and urban communities.[28] For this purpose, Congress appropriated $20 million for each of fiscal years 2006-2008.[29]

[g] State Technologies Advancement Collaborative

In cooperation with states, the DOE must establish a collaborative program for research, development, demonstration, and deployment of technologies in support of common federal and state interests in energy efficiency and renewable and fossil energy.[30] The collaborative program will leverage federal and state funding through cost-shared

---

24 2005 Energy Policy Act § 124(f).

25 2005 Energy Policy Act § 124(b)(3).

26 2005 Energy Policy Act § 125(a).

27 2005 Energy Policy Act § 125(c).

28 2005 Energy Policy Act § 126(a).

29 2005 Energy Policy Act § 126(d).

30 2005 Energy Policy Act § 127(a).

activity, reduce redundancies in federal and state funding, and create relevant multi-state projects.

[h] State Building Energy Efficiency Codes Incentives

In support of Congress' objective in the ECPA to encourage states to update and enforce their energy efficiency codes, Section 128 of the Act provides funds to implement plans to achieve at least 90% compliance with these codes.[31] However, the codes must meet or exceed the 2004 International Energy Conservation Code for residential buildings or the ASHRAE Standard 90.1-2004 for commercial buildings.

[4] Energy Efficient Products

[a] Energy Star Program

Section 131 of the Act amends the EPCA to codify a voluntary program within the DOE and the EPA (the "Energy Star Program") that identifies and promotes energy-efficient products and buildings.[32] The objectives of the program are to reduce energy consumption, improve energy security, and reduce pollution through voluntary labeling of products and buildings that meet the highest energy conservation standards.

[b] HVAC Maintenance Consumer Education Program

Congress amended the EPCA to create a program to be established and administered by the DOE by February 2006 to educate homeowners and small businesses on the energy savings from properly maintained air conditioning, heating, and ventilation systems. The Small Business Administration will also coordinate a government-wide program, based on the Energy Star for Small Business Program, to assist small businesses to become more energy efficient in their operations, to identify financing for energy efficiency upgrades, and to access government procurement opportunities with regard to Energy Star technologies and products. Congress appropriated sums as necessary to accomplish these efforts until expended.[33]

---

[31] 2005 Energy Policy Act § 128.

[32] 2005 Energy Policy Act § 131(a).

[33] 2005 Energy Policy Act § 132.

[c] Public Energy Education Program

The DOE is required to convene a conference by February 2006 to establish a national public energy education program regarding conservation and energy efficiency, the role of energy use in the economy, and the impact of energy use on the environment. The DOE must also provide technical assistance in support of this program as needed.[34]

[d] Energy Efficiency Public Information Initiative

The DOE must carry out an awareness program to inform the public of the need to reduce energy consumption and of the benefits from such reductions.[35] Congress appropriated $90 million for each of fiscal years 2006-2010 for this purpose.[36]

[e] Energy Conservation Standards for Additional Products

Section 135(a) of the Act amends the definition of certain types of lamps under the EPCA, and added definitions for new products.[37] Section 135(b) of the Act requires the DOE to establish testing requirements for refrigerated bottled or canned beverages vending machines by August 8, 2007.[38] Section 135(c), among other things, amends the EPCA to give DOE some discretion to prescribe energy conservation standards or energy use standards for electricity used for the purposes of circulating air through duct work. Section 135(c) of the Act also establishes the standards for fluorescent lamp ballasts.[39]

[f] Energy Conservation Standards for Commercial Equipment

Congress amended the EPCA to create a new definition for very large commercial package air conditioners and heating equipment, automatic commercial ice makers, and commercial cloth washers.[40]

---

[34] 2005 Energy Policy Act § 133(a).

[35] 2005 Energy Policy Act § 134(a).

[36] 2005 Energy Policy Act § 134(e).

[37] 2005 Energy Policy Act § 135(a).

[38] 2005 Energy Policy Act § 135(b).

[39] 2005 Energy Policy Act § 135(c).

[40] 2005 Energy Policy Act § 136(a).

Congress also changed the definitions of small and large commercial package air conditioning and heating equipment.[41] The Act establishes efficiency standards for these products or authorizes the DOE to do so.

[g] Energy Labeling

Section 137 of the Act amends the EPCA to require the DOE or the Federal Trade Commission (FTC), as appropriate, to create labeling requirements for consumer products covered by the EPCA.[42] The FTC must initiate a rulemaking proceeding to determine the effectiveness of these labeling requirements in order to assist consumers in making purchasing decisions and improving energy efficiency.

[h] Intermittent Escalator Study

Section 138 of the Act requires the General Services Administration to conduct a study on the appropriateness of using intermittent escalators throughout the United States.[43]

[i] Energy Efficient Electric and Natural Gas Utility Study

By August 8, 2006, the DOE, in consultation with the National Association of Regulatory Utility Commissioners and the National Association of State Energy Officials, must conduct a study of state and regional policies that promote cost-effective programs to reduce energy consumption, including energy efficiency programs.[44]

[j] Energy Efficiency Pilot Programs

The DOE must establish a pilot program in at least three states but not more than seven states to encourage energy efficiency and to reduce energy consumption.[45] Congress appropriated $5 million for each of fiscal years 2006-2010 to carry out this program.[46]

---

[41] 2005 Energy Policy Act § 136(b).

[42] 2005 Energy Policy Act § 137(a).

[43] 2005 Energy Policy Act § 138(a)–(b).

[44] 2005 Energy Policy Act § 139(a).

[45] 2005 Energy Policy Act § 140(a).

[46] 2005 Energy Policy Act § 140(c).

[k] Report on Failure to Comply With Deadline for New or Revised Energy Conversation Standards

Section 141 of the Act requires the DOE to report to Congress on each deadline in the EPCA that it failed to meet and to provide Congress with reasons for such failure and a plan to remedy it.[47]

[5] Public Housing

[a] Public Housing Capital Fund

Congress expanded the scope of the Capital Fund created under the United States Housing Act of 1937[48] to promote improvement of energy and water-use efficiency by installing fixtures and fittings that conform to the appropriate standards.[49]

[b] Energy Efficient Appliances

Section 152 of the Act requires public housing agencies to purchase energy-efficient appliances that are Energy Star products or FEMP-rated, unless such purchases are not cost-effective to the agency.[50]

[c] Energy Efficiency Standards

Section 153 amends Section 109 of the Cranston-Gonzales National Affordable Housing Act[51] to require the Departments of Agriculture and Housing and Urban Development to develop energy standards and model energy codes by September 30, 2006, for (1) public or assisted housing, (2) single and multifamily residential housing subject to mortgages insured under the National Housing Act or by the Department of Agriculture, and (3) rehabilitation and new construction of public housing under the HOPE VI revitalization grant.[52]

[d] Energy Strategy for HUD

The Department of Housing and Urban Development (HUD) must develop and implement an integrated strategy to reduce utility

---

[47] 2005 Energy Policy Act § 141(a).

[48] 42 U.S.C. § 1437g.

[49] 2005 Energy Policy Act § 151.

[50] 2005 Energy Policy Act § 152.

[51] 42 U.S.C. § 12709.

[52] 2005 Energy Policy Act § 153.

expenses at public and assisted housing facilities through cost-effective energy conservation and efficiency measures.[53] The Act requires HUD to submit a report to Congress on the strategy and to submit an updated report every two years indicating progress in implementing the strategy.

§ 1.02 Title II—Renewable Energy

[1] Scope

Title II of the Act requires dozens of reports and establishes several programs intended to expand the use of renewable resources. Subtitle A includes general initiatives such as an annual Department of Energy (DOE) assessment of renewable energy resources, a federal renewable energy purchase requirement, a photovoltaic energy commercialization program, and various incentives to promote the use of energy efficient practices and the use of renewable energy. Subtitle B amends and updates the treatment of geothermal resource leases under the Geothermal Steam Act of 1970. Subtitle C amends the hydroelectric provisions of the Federal Power Act entitling an applicant or other party to a proceeding to a determination on the record, after opportunity for an agency trial-type hearing of not more than 90 days, on any disputed issues of material fact with respect to hydroelectric license conditions and fishways. Subtitle C also establishes an incentive program for new or expanded hydroelectric capacity so long as it does not necessitate the construction or enlargement of an impoundment or a diversion structure other than for repair or reconstruction. Subtitle D amends existing statutes addressing the energy needs of "insular areas" (such as Puerto Rico, the Virgin Islands and Guam).

[2] General Provisions

[a] Assessment of Renewable Energy Resources

By August 8, 2006, and each year thereafter, the DOE must publish a report containing a detailed inventory of the available amount and characteristics of renewable energy resources within the United States.[1] For the purpose of these reports, renewable energy includes

---

[53] 2005 Energy Policy Act § 154.

[1] 2005 Energy Policy Act § 201(b).

solar, wind, biomass, ocean (including tidal, wave, current and thermal), geothermal, and hydroelectric energy resources.[2] The report also must contain any other information that would be useful in developing renewable energy resources, including (1) descriptions of surrounding terrain, population and load centers, nearby energy infrastructure, location of energy and water resources, and available estimates of the costs needed to develop each resource; (2) an identification of any barriers to providing adequate transmission to deliver renewable energy resources to current and emerging markets; (3) recommendations for removing or addressing such barriers; and (4) ways to provide access to the grid that do not unfairly disadvantage renewable or other energy producers. For this purpose, Congress appropriated $50 million for fiscal years 2006-2010.

[b] Federal Purchase Requirement

The DOE must ensure that, to the extent economically feasible and technically practicable, the total amount of electric energy the federal government consumes during any fiscal year includes the following amounts of renewable energy: not less than 3% in years 2007-2009; not less than 5% in years 2010-2012; and not less than 7.5% in year 2013 and each year thereafter.[3]

In this context, renewable energy means electric energy generated from solar, wind, biomass, landfill gas, ocean (including tidal, wave, current, and thermal), geothermal, municipal solid waste, or new hydroelectric generation capacity achieved from increased efficiency or additions of new capacity at an existing hydroelectric project.[4] If the renewable energy is produced and used on site at a federal facility, or is produced on federal lands and used at a federal facility or is produced on Indian land and used at a federal facility, then the quantity of renewable energy will be considered to be doubled when calculating whether or not the criteria have been met. The DOE must provide a report to Congress on the progress in meeting these goals no later than April 15, 2007, and every two years thereafter.[5]

---

[2] 2005 Energy Policy Act § 201(a).

[3] 2005 Energy Policy Act § 203(a).

[4] 2005 Energy Policy Act § 203(b)(2).

[5] 2005 Energy Policy Act § 203(d).

[c] Use of Photovoltaic Energy in Public Buildings

Congress appropriated $250 million for fiscal years 2006-2010 to establish a photovoltaic energy commercialization program. The program is to accelerate the growth of a commercially viable photovoltaic industry and thereby reduce the consumption of fossil fuels. The program includes a goal of installing solar energy systems in 20,000 federal buildings by the year 2010. Under the program, the General Services Administration (GSA) will provide for the acquisition of photovoltaic solar electric systems and associated storage capability for use in public buildings. In addition, the GSA will establish a program to evaluate photovoltaic solar energy systems and identify steps to ensure that such systems reflect the most advanced technology. An additional $50 million is appropriated to carry out this evaluation program.

[d] Renewable Energy Security

Section 206 of the Act amends the Energy Conservation and Production Act (ECPA), to increase governmental assistance for the weatherization of dwelling units.[6] In addition, the Act revises the District Heating and Cooling Program under the Energy Policy Act of 1992 to provide for a rebate to consumers for expenditures made to install renewable energy systems in dwelling units or small businesses. The amount of the rebate would be the lesser of 25% of the expenditures or $3,000. Congress appropriated $150 million in 2006, $150 million in 2007, $200 million in 2008, $250 million in 2009, and $250 million in 2010 for this provision. Congress also directed the DOE to issue a report by February 8, 2006, containing an inventory of renewable fuels available for consumers and a projection of future inventories of renewable fuels based on these incentives.[7]

[e] Additional Incentives Promoting the Use of Renewable Energy

Section 209 of the Act establishes grants to encourage rural and remote communities to increase energy efficiency, including a preference for renewable energy facilities. Congress appropriated $140 million for the years 2006-2012 for this program. Section 210 of the

---

6 2005 Energy Policy Act § 206(a); 42 U.S.C. § 6865(c).

7 2005 Energy Policy Act § 206(d).

Act establishes grants to improve the commercial value of forest biomass for electric energy, useful heat, transportation fuels, and other commercial purposes. Congress appropriated $550 million for fiscal years 2006-2016 for these grants.

[3] Geothermal Energy

Subtitle B of Title II amends the existing statutes related to geothermal energy and geothermal steam. The provisions amend the leasing and lease procedures for geothermal resources, as well as the royalty system for federal lands, to encourage the use of geothermal resources.

[a] Competitive Leasing

The Act makes most land that can be leased under the Geothermal Steam Act of 1970[8] subject to a competitive leasing process. However, land can be leased through a noncompetitive process if the Secretary of the Interior has not received any bids in a competitive lease sale. In addition, lands that are subject to a mining claim for which a plan of operations has been approved by the relevant federal land management agency may qualify for noncompetitive leasing through the mining claim holder.[9]

[b] Fees and Leasing for Direct Use

The Secretary of the Interior must establish a schedule of fees in lieu of royalties for geothermal resources where the lessee or its affiliate leases the land for the direct use of the geothermal resources without sale and for purposes other than commercial generation of electricity.[10] The schedule of fees may be based on the quantity or the thermal content, or both, of the geothermal resources to be used, must ensure a fair return to United States for use of the resource, and must encourage development of the resource. If the lessee is a state, tribal, or local government and uses the land consistent with this provision, then the Secretary must charge only a nominal fee.

In addition, the Secretary may exempt direct use leases from the competitive bidding process if the Secretary determines that no competitive interest in the lease exists.

[8] 30 U.S.C. § 1003.

[9] 2005 Energy Policy Act § 222.

[10] 2005 Energy Policy Act § 223(a).

[c] Royalties and New-Term Production Incentives

Congress reduced the level of rents and royalties applicable to the use of federal land for purposes of geothermal resources. The preexisting legislation required a royalty of not less than 10% nor more than 15% applicable to the value of the steam heat or energy derived from the geothermal resource.[11] Under Section 224 of the Act, the royalty on electricity produced using geothermal resources, other than direct use of geothermal resources, shall be not less than 1% and not more than 2.5% of the gross proceeds from the sale of electricity produced from such resources during the first 10 years of production under the lease and not less than 2% and not more than 5% of the gross proceeds after such 10-year period.[12] In addition, the Secretary of the Interior may provide to a lessee a credit against the royalties owed under the Geothermal Steam Act in an amount equal to the value of electricity provided under contract to a state or county government that is entitled to a portion of such royalties under Section 20 of the Geothermal Steam Act. However, the Secretary must first approve the contract between the lessee and the state or county government for such in-kind payment. The contract must establish a specific method to determine the value of the credits and the maximum credit must equal the royalty value owed to the state or county.[13]

In Section 224(c) of the Act, Congress also created an incentive to increase the near-term production of geothermal energy from existing leases. Congress directed that the royalty required to be paid shall only be 50% of the otherwise applicable royalty fee on any lease issued before the date of enactment of the Act, unless the lease is converted to new terms that (1) achieve commercial production of energy from a facility within the six-year period starting on the Act's enactment, or (2) are qualified expansion geothermal energy. This incentive will apply only to new commercial production of energy from a facility in the first four years of such production.[14] Qualified expansion geothermal energy is geothermal energy produced from a generation facility that achieves a production increase of more than

11 30 U.S.C. § 1004(a).

12 2005 Energy Policy Act § 224(a).

13 2005 Energy Policy Act § 224(a).

14 2005 Energy Policy Act § 224(c).

10% as a result of expansion carried out during the six year period beginning on the date of the Act's enactment where such production increases are greater than 10% of the average production of the facility during the five-year period proceeding the facility's expansion.[15] Such lessees have the option to confirm that the lease qualifies under pre-existing provisions of the Geothermal Steam Act and that the royalties should be based on those provisions of existing law.[16] These requests must be submitted to the Secretary of the Interior pursuant to regulations that must be issued by the Secretary.

[d] Coordination of Geothermal Leasing and Permitting on Federal Land

To streamline the leasing and permitting of geothermal resources on certain federal lands, the Secretary of the Interior and the Secretary of Agriculture must enter into, and submit to Congress, a Memorandum of Understanding (MOU) that coordinates the leasing and permitting process for geothermal development on public lands and national forest system lands under their respective jurisdictions.[17] This MOU must establish a procedure to process geothermal lease applications including lines of authority, steps in the application process and time limits for application processing. The MOU also must establish a five-year program for geothermal leasing of lands in the national forest system and include a process for updating the program every five years. In addition, the MOU must include a program for reducing the backlog of geothermal lease applications pending on January 1, 2005. The procedures applicable to the duration of leases and the level of work commitment that must be shown by the leaseholder have been revised to encourage diligent development of the lease.[18]

[e] Advanced Royalties Required for Cessation of Production

If production under a lease ceases after commercial production, the lease will remain in effect for a period of not more than an aggregate of 10 years beginning on the date production ceases so long as, during

[15] 2005 Energy Policy Act § 224(d).

[16] 2005 Energy Policy Act § 224(e).

[17] 2005 Energy Policy Act § 225(a).

[18] S. Rep. No. 109-78, at 22 (2005).

the period of non-production, the lessee pays royalties in advance at the monthly average rate at which the royalties were paid during the period of production.[19]

[f] Annual Rents

Congress also increased the annual rental rate to be paid on federal geothermal leases. For noncompetitive lease sales, the initial rate is $1 per acre or fraction thereof and in the case of a competitive lease sale, the rate is $2 per acre or fraction thereof and $3 per acre for each of the second through tenth years. For each year after the tenth year, the annual rental lease payment is to be $5 per acre. The Secretary of the Interior must terminate any lease where the rental is not paid in accordance with the Act.[20]

[4] Hydroelectric

[a] Alternative Conditions and Fishways

Part I of the Federal Power Act (FPA) governs the issuance of licenses for the construction, operation, and maintenance of hydroelectric generation facilities by private entities, cooperatives or non-federal governmental authorities.[21] Section 10 of the FPA charges the Federal Energy Regulatory Commission (FERC) with the responsibility for issuing licenses that "in the judgment of the Commission will be best adapted to a comprehensive plan for improving or developing a waterway."[22] However, the statute also reserves certain authority to other federal departments to issue terms and conditions that the Commission must include as part of any license that it might issue.

Where a project occupies public lands or reservations of the United States, Section 4(e) of the FPA provides that the license "shall be subject to and contain such conditions as the Secretary of the department under whose supervision such reservation falls shall deem necessary for the adequate protection and utilization of such reservation."[23] Any license must also provide for the construction,

---

[19] 2005 Energy Policy Act § 232.

[20] 2005 Energy Policy Act § 233.

[21] 16 U.S.C. § 791a *et seq.*

[22] 16 U.S.C. § 803(a)(1).

[23] 16 U.S.C. § 797(e).

maintenance, and operation by the licensee, at its own expense, of fishways that may be prescribed by the appropriate federal department under Section 18 of the FPA.[24] As the FERC may not otherwise modify the prescriptions rendered under Sections 4(e) and 18, any attempts to influence the fashioning of such provisions at the administrative level must be conducted before the Department vested with the authority to issue the prescription. Section 241 of the Act amends both Section 4(e) and Section 18 of the FPA to provide for trial-type hearings before the Department with the prescriptive authority and by allowing for the consideration of alternative conditions and prescriptions.

Subsections (a) and (b) of Section 241 of the Act provide for the trial-type hearing by amending, respectively, both Section 4(e) and Section 18 of the FPA to allow the license applicant or any party to the licensing proceeding to have a determination on the record, after a hearing of no more than 90 days, on any issue of material fact, with respect to the conditions proposed either for the adequate protection and utilization of the federal reservation covered by Section 4(e) or the fishway addressed by Section 18. In an effort to secure prompt implementation of this new opportunity for a hearing, Section 241 directs the Secretaries of the Interior, Commerce, and Agriculture to establish jointly, by rule and in consultation with the FERC, within 90 days of the enactment of the Act, the procedures for expedited hearings, including the opportunity to undertake discovery and cross-examine witnesses.[25]

Section 241(c) of the Act creates a new Section 33 of the FPA which opens up proceedings under Sections 4(e) and 18 of the FPA for the consideration of alternatives to the conditions and prescriptions being proposed by the various departmental Secretaries. Where a Secretary of a department with supervisory authority over the reservation deems a condition necessary under Section 4(e) of the FPA, the license

---

[24] 16 U.S.C. § 811. Congress has specifically declared that a fishway must provide for both upstream and downstream passage of fish. Section 1701(b) of the Energy Policy Act of 1992. 106 Stat. 2776, 3008 (1992).

[25] 2005 Energy Policy Act § 241(a) & (b). In timely compliance with the Act, the Secretaries of Agriculture, Interior and Commerce jointly published interim final rules establishing the procedures. Resource Agency Procedures for Conditions and Prescriptions in Hydropower Licenses, 70 Fed. Reg. 69,840 (Nov. 17, 2005) (to be codified at 50 C.F.R. Part 221).

applicant or any party to the proceeding may propose an alternative condition which the Secretary must accept if, based on substantial evidence, the alternative (1) provides for the adequate protection and use of the reservation, and (2) will either, as compared to the condition initially proposed by the Secretary, cost significantly less to implement or result in improved operation of the project works for electricity production. The Secretary must consider evidence provided for the record by any party to the licensing proceeding, or otherwise available to the Secretary, including any evidence provided by the FERC, on the cost of implementation or operational impacts for the production of electricity associated with the proposed alternative. A written statement from the Secretary must accompany any condition under Section 4(e) or alternative condition adopted under the new Section 33. That statement must explain the basis for the conditions, and the reason for not accepting any alternative condition. Further, the Secretary is charged with demonstrating that equal consideration has been given to the effects of the condition adopted and alternatives on energy supply, distribution, cost, and use, flood control, navigation, water supply, air quality, and the preservation of other aspects of environmental quality. All studies, data, and other factual information available to the Secretary and relevant to the decision must also be submitted to the FERC, along with the statement.

[b] Hydroelectric Production Incentives

Congress provided for incentives designed to encourage new investment in hydroelectric generating facilities and increase the amount of the Nation's energy derived from the use of falling water. These measures include incentives for increased generating capacity and efficiency improvements for non-federal hydroelectric projects.

Section 242 of the Act provides that, where a turbine or other generating device owned or solely operated by a non-federal entity is added to an existing dam or conduit, the Secretary of Energy, acting upon an application, is authorized to make incentive payments of 1.8 cents per kwh based on the number of kilowatt hours of hydroelectric energy generated by the qualified facility, provided that the facility commences operation during the period of 10 fiscal years beginning with the first full fiscal year occurring after the enactment of the Act.[26]

---

[26] 2005 Energy Policy Act § 242.

Once the facility has commenced operation, it is eligible to receive payments for a period of 10 fiscal years, and no more, beginning with the fiscal year in which electric energy generated from the facility is first eligible for such payments. The incentive payments, which Congress limited to no more than $750,000 in any one calendar year for any one qualifying facility, are subject to the availability of funds appropriated for that year. The amounts stipulated by the statute for both the per kwh charge and total payments are subject to adjustment for inflation. To fund this program, Congress authorized $10 million to be appropriated for each of the fiscal years 2006-2015.

Where the installation of new turbine or other generating equipment may not be an option, Section 243 of the Act authorizes the Secretary of Energy to make incentive payments for capital expenditures at existing hydroelectric facilities related to improving the efficiency of such facilities by at least 3%. The incentive payments may not exceed 10% of the costs of a capital improvement and not more than one payment may be made with respect to improvements at a single facility, with any such payment being limited to $750,000 for such a facility. To fund these incentive payments, Congress authorized not more than $10 million to be appropriated for each of the fiscal years 2006-2015. While Section 242 governing production incentives clearly defined the terms of eligibility, Section 243 lacks any such provision for the capital improvements. Section 243 also does not allow for any adjustment to the cap on the incentive payments to accommodate any inflation in costs.

[c] Alaska State Jurisdiction Over Small Hydroelectric Projects

Under the outstanding licensing jurisdiction over small hydroelectric projects bestowed on the State of Alaska by Section 32 of the FPA, the state is required to include in any license it issues the necessary provisions for the protection and mitigation of fish and wildlife resources based on conditions received pursuant to the Fish and Wildlife Coordination Act[27] from the National Marine Fisheries Service, the United States Fish and Wildlife Service, and state resource agencies for qualifying projects located within Alaska. Section 244 of the Act amends Section 32(a)(3)(c) of the FPA to authorize the

[27] 16 U.S.C. § 661 *et seq.*

State of Alaska to decline to include a recommended condition, if the state determines that the recommendation is inconsistent with the standards of Sections 32(a)(1) and (2) of the FPA that require as part of the state's hydro regulatory program that the program protect the public interest, including protection of the environment to the same extent that the Commission must do under Part I of the FPA, and that the program give equal consideration to certain stipulated issues.[28] Should Alaska decline to accept a recommended condition, the new Section 32(j) requires that any such rejection must be in accordance with the procedures set forth in Section 10(j)(2) of the FPA.[29]

§ 1.03 Title III—Oil and Gas

[1] Scope

Title III of the 2005 Energy Policy Act contains a wide-ranging mix of initiatives designed to ensure an adequate and growing supply of natural gas and oil resources for the United States. Of particular note are the Act's modifications to Section 3 of the Natural Gas Act (NGA),[1] which make clear that the Federal Energy Regulatory Commission (FERC) is the lead agency in making determinations related to the siting and approval of liquefied natural gas (LNG) import terminals, as well as other modifications that adopt a variety of

---

[28] The list of issues are (1) energy conservation; (2) the protection, mitigation of damage to, and enhancement of, fish and wildlife (including related spawning grounds and habitat); (3) the protection of recreational opportunities; (4) the preservation of other aspects of environmental quality; (5) the interests of Alaska Natives; and (6) other beneficial public uses, including irrigation, flood control, water supply, and navigation. 16 U.S.C. § 823c(a)(2).

[29] Section 10(j)(2) of the FPA, 16 U.S.C. § 803(j)(2), states in part:

> Whenever the Commission believes that any recommendation referred to in [Section 10(j)(1)] may be inconsistent with the purposes and requirements of this Part or other applicable law, the Commission and agencies referred to in paragraph (1) shall attempt to resolve any such inconsistency, giving due weight to the recommendations, expertise, and statutory responsibilities of such agencies. If, after such attempt, the Commission does not adopt in whole or in part a recommendation of any such agency, the Commission shall publish each of the following findings (together with a statement of the basis for each of the findings):
>
> (A) A finding that adoption of such recommendation is inconsistent with the purposes of and requirements of this Part or with other applicable provisions of law.
>
> (B) A finding that the conditions selected by the Commission comply with the requirements of [Section 10(j)(1)].

[1] 15 U.S.C. § 717b.

production incentives designed to foster production from marginal or difficult-to-access gas and oil producing properties. In recognition of the energy market manipulation and deceptive practices that occurred in the early part of this millennium, Congress has substantially enhanced the FERC's oversight of the natural gas marketing and trading business, and has given the FERC significantly increased penalty authority for violations of the NGA and the Natural Gas Policy Act (NGPA). Additionally, the Act provides the framework for the FERC to issue new rules to deter and penalize market manipulation and to promote natural gas market transparency—an initiative that the FERC has seized.

[2] Petroleum Reserve and Home Heating Oil

Section 301 of the Act contains a number of ministerial changes to the Energy Policy and Conservation Act (EPCA).[2] Most of the changes ensure that the existing law related to Domestic Supply Availability, including Domestic Supply,[3] Strategic Petroleum Reserve,[4] Authority to Contract for Petroleum Product Not Owned by the United States,[5] and Northeast Home Heating Oil Reserve,[6] remains in force and viable. One provision contained within Section 301 worth specific mention requires the Secretary of Energy, "as expeditiously as practicable, without incurring excessive cost or appreciably affecting the price of petroleum products to consumers," to "acquire petroleum in quantities sufficient to fill the Strategic Petroleum Reserve to the 1,000,000,000-barrel capacity" that previously had been authorized. The Secretary of Energy is required to "develop, with public notice and opportunity for comment, procedures consistent with the objectives of this section to acquire petroleum for the Reserve." Section 303 of the Act further requires the Secretary to complete a proceeding, no later than one year after the date of enactment of the Act, to select from previously considered sites, sites that would allow the Secretary to acquire the full 1,000,000,000 barrels.

---

[2] 42 U.S.C. § 6201 *et seq.*

[3] 42 U.S.C. §§ 6212–6217.

[4] 42 U.S.C. §§ 6231–6247b.

[5] 42 U.S.C. §§ 6249–6249c.

[6] 42 U.S.C. §§ 6250–6250e.

[3] Natural Gas

[a] LNG Import Facility Permitting, Construction, and Regulation

The Act includes a number of changes that are designed to ensure that the FERC is the lead agency with respect to the review and approval of applications for LNG import terminals. In particular, it resolves any question that the FERC has the authority under the NGA to review and approve applications for authority to construct LNG terminals. While the clear message from the Act is that the FERC will lead the way on LNG permitting, there are a number of provisions that ensure that states play a role in the process. Importantly, the Act does not remove a state's existing authority to implement its delegated federal authority under the Clean Water Act,[7] the Clean Air Act,[8] and the Coastal Zone Management Act.[9] Thus, a state retains significant oversight over the approval of LNG terminals through these federal laws.[10] What remains less clear, however, is whether the provisions giving the FERC exclusive authority over the approval of LNG terminals will lead to new challenges claiming that a state or local law used to block the construction of an LNG terminal is preempted by the FERC's NGA authority over LNG import terminals.[11]

---

[7] 33 U.S.C. §§ 1251–1387.

[8] 42 U.S.C. §§ 7401–7671q.

[9] 33 U.S.C. §§ 1451–1465.

[10] *See* Joseph T. Kelliher, Chairman, Federal Energy Regulatory Commission, Editorial, *Energy Bill Expands State Inspection Powers*, Wall St. J., Aug. 8, 2005, at A11 ("In addition to applicable state and local laws, the states have federal authority delegated to them under the Coastal Zone Management Act, the Clean Air Act and the Clean Water Act. These authorities remain unchanged, and under any of those federal statutes, the states can effectively veto a proposed LNG facility regardless of any FERC decision under the Natural Gas Act.").

[11] *See* Schneidewind v. ANR Pipeline Co., 485 U.S. 293, 300 (1988); Transcontinental Gas Pipe Line Corp. v. State Oil & Gas Bd., 474 U.S. 409, 425 (1986); Northern Natural Gas Co. v. State Corp. Comm'n, 372 U.S. 84, 89 (1963); Maryland v. Louisiana, 451 U.S. 725, 749-752 (1981); Exxon Corp. v. Eagerton, 462 U.S. 176, 184 (1983).

[b] No Open Access for LNG Terminals—Codification of the FERC's Hackberry Policy

The FERC significantly shifted its policies related to LNG import terminals when it issued its decision in *Hackberry LNG Terminal, L.L.C.* (*Hackberry*).[12] In *Hackberry*, the FERC determined that an applicant for a new LNG terminal import terminal could operate that terminal on a non-open-access, proprietary basis, thereby differentiating the FERC's regulation of LNG import terminals from its regulation of interstate natural gas pipelines.[13] While interstate pipelines are required to operate on an open-access basis, subject to extensive FERC regulation of rates, as well as terms and conditions of service, in *Hackberry*, the FERC determined that the project developer did not need to provide open-access to its new LNG terminal and determined that there was no "regulatory need to require a tariff and a rate schedule as a condition of approving the LNG terminal under Section 3."[14] The FERC determined that this policy shift could "provide incentives to develop additional energy infrastructure to increase much needed [gas] supply into the United States, while at the same time ensuring competitive commodity prices and an open-access interstate pipeline grid."[15] After *Hackberry*, the FERC consistently followed the non-open-access approach for the approval of new LNG import facilities.[16]

The Act codified the *Hackberry* policy. Section 311(3)(B)(ii) prohibits the FERC, before January 1, 2015, from conditioning an order on (1) "a requirement that the LNG terminal offer service to customers other than applicant, or any affiliate of the applicant"; (2) "any regulation of the rates, charges, terms, or conditions of service of the LNG terminal"; and (3) "a requirement to file with the [FERC]

---

[12] Hackberry LNG Terminal, L.L.C., 101 FERC ¶ 61,294 (2002), *order issuing certificates & on reh'g*, 104 FERC ¶ 61,269 (2003).

[13] Hackberry LNG Terminal, L.L.C., 101 FERC ¶ 61,294, at P 3 (2002), *order issuing certificates & on reh'g*, 104 FERC ¶ 61,269 (2003).

[14] Hackberry LNG Terminal, L.L.C., 101 FERC ¶ 61,294, at P 3, 26 (2002), *order issuing certificates & on reh'g*, 104 FERC ¶ 61,269 (2003).

[15] Hackberry LNG Terminal, L.L.C., 101 FERC ¶ 61,294, at P 3, 26 (2002), *order issuing certificates & on reh'g*, 104 FERC ¶ 61,269 (2003).

[16] *See, e.g.*, Freeport LNG Development, L.P., 107 FERC ¶ 61,278 (2004); Sabine Pass LNG, L.P., 109 FERC ¶ 61,324 (2004); Corpus Christi LNG, L.P., 111 FERC ¶ 61,081 (2005); Golden Pass Terminal LP, 112 FERC ¶ 61,041 (2005); Ingleside Energy Center, LLC, 112 FERC ¶ 61,101 (2005).

schedules or contracts related to the rates, charges, terms, or conditions of service of the LNG terminal." This provision terminates on January 1, 2030.[17]

The Act additionally recognized that there are existing LNG import terminals that offer open access service under prior FERC orders.[18] The Act ensures that existing customers of those open-access LNG import terminals will not be adversely affected by any order related to an expansion, modification, or change to an existing open-access facility. The Act additionally requires the FERC to enter into a Memorandum of Understanding with the Secretary of Defense to coordinate review of any LNG facility that may affect an active military installation and to obtain concurrence of the Secretary of Defense if any LNG authorization would affect the training or activities of an active military installation.[19]

[c] FERC Exclusive Jurisdiction Over Onshore LNG Terminal Siting

When the FERC issued its *Hackberry* decision, no one questioned whether the FERC was the agency empowered to consider and rule on appropriateness of the siting and regulation of LNG import facilities that are onshore or in state waters.[20] After *Hackberry*, a case involving a proposed LNG import terminal in California called the FERC's regulatory primacy into question. Sound Energy Solutions (SES) filed an application with the FERC under Section 3 of the NGA for authority to site, construct, and operate an LNG terminal in Long Beach, California. The California Public Utilities Commission (CPUC) intervened and lodged a protest before the FERC, arguing that SES had

---

17 2005 Energy Policy Act § 311(e)(3)(C).

18 Those facilities are the Cove Point terminal in Lusby, Maryland; the Lake Charles terminal in Lake Charles, Louisiana; and the Elba Island terminal in Savannah, Georgia. The fourth terminal subject to the FERC's regulation is the Everett terminal in Everett, Massachusetts (which, although operating under a FERC tariff, is not open access).

19 2005 Energy Policy Act § 311(f)(2)–(3).

20 The regulation of LNG import facilities located offshore and beyond state waters is covered by the Deepwater Port Act of 1974 (DPA), as amended by the Maritime Transportation Security Act of 2002. 33 U.S.C. § 1503 (2005). Regulatory authority under the DPA is shared by the United States Department of Transportation's Maritime Administration and the U.S. Coast Guard, which is now part of the Department of Homeland Security.

to obtain a certificate of public convenience and necessity from the State of California before beginning to build its terminal. The FERC rejected that argument, citing previous case law and its own regulations, and pledged to "exercise its flexibility under NGA Section 3 to regulate the LNG import terminal as well as the pipeline facilities that will deliver gas into state regulated facilities downstream," and concluded that "the LNG facilities proposed by SES cannot be regulated by the State."[21] The CPUC ultimately appealed that case to the United States Court of Appeals for the Ninth Circuit. At the time that the Act was passed, the SES appeal before the Ninth Circuit was pending.

The Act ended that uncertainty, as intended.[22] The CPUC withdrew its appeal.[23] Section 311(e)(1) made clear that "[t]he [FERC] shall have the exclusive authority to approve or deny an application for the siting, construction, expansion, or operation of an LNG terminal." Nevertheless, Congress did attempt to create a procedural vehicle that would allow the states and localities to raise issues regarding LNG terminal siting.

---

21 Sound Energy Solutions, 106 FERC ¶ 61,279, at P 26 (2002).

22 *See* The Energy Policy Act of 2005: Hearings Before the Subcommittee on Energy and Air Quality of the House of Representatives Committee on Energy and Commerce, 109th Cong. 31 (2005) (statement of Cynthia A. Marlette, General Counsel, Federal Energy Regulatory Commission) ("The Commission currently is involved in litigation in the U.S. Court of Appeals for the 9th Circuit with respect to the scope of its authority to site LNG terminal facilities. Legislation could end regulatory uncertainty by clarifying the Commission's authority in this area. A single federal agency should have the statutory authority to determine whether a specific proposal for LNG infrastructure development is in the public interest."); Joseph T. Kelliher, Chairman, Federal Energy Regulatory Commission, Editorial, *Energy Bill Expands State Inspection Powers*, Wall St. J., Aug. 8, 2005, at A11 ("What the energy bill does is affirm the Federal Energy Regulatory Commission's long-standing exclusive Natural Gas Act authority to authorize LNG import terminals. This was challenged by the California Public Utilities Commission in a case now pending before the Ninth Circuit. This lawsuit threatened to bring the development of LNG import facilities to a standstill, which is why Congress acted. By affirming FERC's authority under the Natural Gas Act, the legislation resolves the regulatory uncertainty created by the commission's challenge to FERC's established authority. In short, the new law maintains the status quo that existed before the legal challenge.").

23 Californians for Renewable Energy, Inc. v. FERC, Nos. 04-73650, 04-75240 (9th Cir. Oct. 6, 2005) (order construing CPUC response as motion for voluntary dismissal and dismissing).

[d] Codification of the NEPA Pre-Filing Process and State Advisory Role

Prior to the Act, the FERC had implemented a voluntary, consultative process with LNG terminal applicants (as well as companies planning to build other natural gas infrastructure)[24] called the National Environmental Policy Act (NEPA) pre-filing process. The purpose of the NEPA pre-filing process was to provide for early FERC staff and other governmental and non-governmental stakeholder involvement in the application process before a project sponsor filed a formal application with the FERC requesting authority to construct an LNG terminal or other natural gas project. The pre-filing process included the filing of drafts of the environmental resource reports, which accompany any application for authority to construct an LNG import facility, public outreach, and scoping meetings led by the FERC staff. This open process was expected to provide for open lines of communication between all who are interested in a project before an application is filed, thereby locking in the parameters of a project.

The Act codified the NEPA pre-filing process, required applicants for LNG terminals to comply with the process, and provided that the process must begin at least six months before a formal application is filed.[25] The Act stated that the FERC should "encourage applicants to cooperate with State and local officials."[26] The FERC issued a final rule revising its regulations to implement the NEPA pre-filing process on October 7, 2005.[27] Revised 18 C.F.R. § 157.21 provides a detailed description of the pre-filing process and the various requirements and milestones associated with the process.

---

[24] *See* 18 C.F.R. § 157.22 (2005) (providing procedures for pre-filing process under Section 7 of the Natural Gas Act); *see also* 18 C.F.R. § 153.12 (2005) (noting that the procedures included in Section 157.22 also would apply to Section 3 applicants); *Regulations Implementing Energy Policy Act of 2005; Pre-Filing Procedures for Review of LNG Terminals and Other Natural Gas Facilities*, 113 FERC ¶ 61,015, at P 3 n.4 (2005) (*Pre-Filing Rule*) (noting that FERC had implemented additional staff guidance to make the process workable in past).

[25] 2005 Energy Policy Act § 311(d); NGA § 311(c)(3A)(a).

[26] 2005 Energy Policy Act § 311(d); NGA § 311(c)(3A)(a).

[27] *Pre-Filing Rule*, 113 FERC ¶ 61,015. The regulations describing the procedure are now included in 18 C.F.R. § 157.21 (applying specifically to NGA § 7 projects); 18 C.F.R. § 153.12 notes that the regulations included in § 157.21 apply to projects under Section 3, such as LNG terminals.

[e] State Advisory Role

Much of the debate surrounding what now has become Section 311 of the Act focused on the appropriate role for states and localities in the LNG import terminal approval process. Issues related to safety and security of LNG terminals have taken center stage since the terrorist activities that took place on 9/11. In addition, politicians on the federal, state, and local levels began to conclude that LNG import facilities, as well as LNG tankers, were significant draws to terrorists. As a result, Congress sought to ensure that state and local governments would have a meaningful say in the LNG terminal approval process.

Therefore, Section 311(d) of the Act provides a formal advisory role to states facing the installation of LNG import facilities. The governor of a state where an LNG terminal is proposed must designate a state agency "for purposes of consulting with the FERC regarding an application under section 3." It then requires that the FERC "consult with such State agency regarding State and local safety considerations prior to issuing an order pursuant to section 3."[28] Each state agency also is provided the right to furnish "an advisory report on State and local safety considerations to the [FERC] with respect to an application no later than 30 days after the application was filed with the [FERC]."[29] The FERC does not have to take any particular action regarding the contents of that advisory report, except to "review and respond specifically to the issues raised by the State agency" regarding the state and local safety considerations outlined in the Act.[30] The Act further allows a state commission to undertake safety inspections after a terminal is operational as long as the state commission follows federal regulations and guidelines. The state agency then may transmit any safety concerns to the FERC, which, in turn, must pass information about the alleged safety violations to other appropriate federal agencies.[31] The appropriate federal agency must then take the action it

[28] State and local safety considerations include: "(1) the kind and use of the facility; (2) the existing and projected population and demographic characteristics of the location; (3) the existing and proposed land use near the location; (4) the natural and physical aspects of the location; (5) the emergency response capabilities near the facility location; and (6) the need to encourage remote siting." 2005 Energy Policy Act § 311(d).

[29] 2005 Energy Policy Act § 311(d).

[30] 2005 Energy Policy Act § 311(d).

[31] 2005 Energy Policy Act § 311(d).

considers appropriate and must notify the state commission of that action.[32]

[f] Development of Emergency Response Plan and Cost Sharing Plan

In a further recognition of the importance that safety issues have taken in the LNG terminal discourse, the Act requires the FERC to require LNG terminal operators to develop an "Emergency Response Plan" in consultation with the U.S. Coast Guard and state and local agencies.[33] The FERC must approve the Emergency Response Plan before any final approval to begin construction. Thus, the Emergency Response Plan is not a prerequisite to receiving approval from the FERC for authorization under Section 3.[34] The Emergency Response Plan must include a "cost-sharing plan," which must "include a description of any direct cost reimbursements that the applicant agrees to provide to any state and local agencies with responsibility for security and safety" at the LNG terminal or near vessels that serve the terminal.[35] The Act does not provide clear guidelines or requirements for the costs a terminal operator will cover.

[g] Procedural Rules and Oversight

The Act formally designates the FERC as "lead agency with regard to review and approval of LNG terminals."[36] This designation has two components. The first is one familiar to many practitioners: the FERC will be the lead agency for purposes of coordinating and complying with the NEPA.[37] This designation puts into law what had been the established practice. The second is an interesting new procedural vehicle whose importance will only be known after seeing it in practice. The Act provides that "[e]ach Federal and State agency considering an aspect of an application for Federal authorization shall cooperate with the [FERC] and comply with the deadlines established by the [FERC]."[38] The FERC is expressly given the authority to set

32 2005 Energy Policy Act § 311(d).

33 2005 Energy Policy Act § 311(e)(1).

34 2005 Energy Policy Act § 311(e)(1).

35 2005 Energy Policy Act § 311(e)(2).

36 2005 Energy Policy Act § 313(a).

37 2005 Energy Policy Act § 313(a).

38 2005 Energy Policy Act § 313(a).

the schedule for all federal authorizations, including those federal authorizations that are delegated to a state administrative agency. If any federal or state agency acting under delegated federal authority fails to meet the deadlines for completion of a proceeding covered by the FERC's schedule, a project applicant is permitted to bring an action in the United States Court of Appeals for the District of Columbia Circuit to address the agency delay and to seek an order from the court to require the delinquent agency to meet a court-imposed schedule.[39]

The Act also provides that any challenge of a FERC action under Sections 3 and 7 of the NGA, or the action of any federal or state agency acting under delegated federal authority, must be brought in the U.S. Court of Appeals for the circuit where the facility is located.[40] Thus, the appropriate court for merits-based appeals is different than the appropriate court for raising a failure to meet the FERC-imposed schedule and deadlines. To facilitate judicial review, the Act requires the FERC to maintain a consolidated record of all decisions made or actions taken by all federal agencies and state agencies acting under federally delegated authority.[41] The consolidated record will be the record for all appeals or reviews under the Coastal Zone Management Act of 1972,[42] and for appeals or other review under the provisions established in the Act.

[h] Federal and State Natural Gas Forums

Section 317 of the Act requires that the Secretary of Energy, in cooperation and consultation with the Secretary of Transportation, the Secretary of Homeland Security, the FERC, and the governors of coastal states, convene at least three forums on LNG. The forums must be in areas where LNG terminals are under consideration. The purpose of the forums is to provide at a minimum a chance to provide public education on the role of LNG in meeting supply needs, the regulatory processes at the federal and state level related to the approval of LNG

---

[39] 2005 Energy Policy Act § 313(a) (providing for new NGA provisions included at 15 U.S.C. § 717n(d)(2)–(3)).

[40] 2005 Energy Policy Act § 313(a) (providing for new NGA provisions included at 15 U.S.C. § 717n(d)(1)).

[41] 2005 Energy Policy Act § 313(d).

[42] 16 U.S.C. § 1451.

terminals, environmental requirements associated with the permitting of LNG terminals, and safety and security measures.

[i] Market-Based Rates for New Natural Gas Storage Facilities

Section 312 of the Act allows the FERC to take a light-handed approach to the regulation of the rates for new natural gas storage facilities. In the past, the FERC had been willing to consider and approve the use of market-based rates for new storage facilities, but that decision was rooted in the FERC's traditional regulatory approach and required applicants to prepare detailed market power analyses.[43]

Section 312 amends Section 7 of the NGA to expressly permit the FERC to approve the use of market-based rates for new natural gas storage projects, even if an applicant cannot prove that it lacks market power. To do so, the FERC must determine that "market-based rates are in the public interest and necessary to encourage the construction of the storage capacity in the area needing storage services" and that "customers are adequately protected."

[j] Gas Trading Regulation and Market Manipulation

In reaction to Enron, the meltdown of the California market, and other competitive abuses in the industry, Congress gave the FERC enhanced authority to monitor, deter and penalize improper behavior in the natural gas industry.[44] Section 315 provides FERC with additional tools to prevent market manipulation.[45] Section 316 encourages price transparency in markets for the sale or transportation of physical natural gas in interstate commerce.[46] And, Sections 314 and 318 provide enhanced civil and criminal penalties.[47]

[i] Market Manipulation

Section 315 amends the NGA by adding Section 4A,[48] which makes it unlawful for an entity to directly or indirectly use, in connection

---

[43] *E.g.,* Starks Gas Storage L.L.C., 111 FERC ¶ 61,105, at P 35, *reh'g denied,* 111 FERC ¶ 61,484 (2005).

[44] 2005 Energy Policy Act §§ 314, 315, 316, 318.

[45] 2005 Energy Policy Act § 315.

[46] 2005 Energy Policy Act § 316.

[47] 2005 Energy Policy Act §§ 314, 318.

[48] 15 U.S.C. § 717c-1.

with the purchase or sale of natural gas or gas transportation subject to the FERC's jurisdiction, "any manipulative or deceptive device or contrivance" (as those terms are used and defined by Section 10(b) of the Securities Exchange Act of 1934)[49] in contravention of the FERC's rules and regulations.[50] Section 315 is intended to enhance the FERC's authority to prohibit manipulation of the energy markets and, coupled with the expanded civil penalty authority, Section 315 provides the FERC with more effective tools to assure workable competitive markets.[51] The Act does not create a private right of action.[52]

The FERC issued a Notice of Proposed Rulemaking (NOPR) setting forth proposed regulations on October 20, 2005.[53] Under one of the FERC's proposed regulations, it will be unlawful for any entity, directly or indirectly, in connection with the purchase or sale of natural gas or the purchase or sale of transportation services subject to the FERC's jurisdiction, (1) to use or employ any device, scheme, or artifice to defraud; (2) to make any untrue statement of a material fact or to omit to state a material fact necessary in order to make the statements made, in the light of the circumstances under which they were made, not misleading; or (3) to engage in any act, practice, or course of business that operates or would operate as a fraud or deceit upon any person.[54] The proposed regulation states that nothing in it shall be construed to create a private right of action.[55]

Section 315 applies to "any entity." Therefore, the FERC's proposed regulation applies not just to jurisdictional market-based rate sellers, natural gas pipelines, or holders of blanket certificate authority, but also to governmental utilities and other market participants.[56] Thus,

---

49 15 U.S.C. § 78j(b).

50 2005 Energy Policy Act § 315.

51 Prohibition of Energy Market Manipulation, 70 Fed. Reg. 61,930, 61,931 (Oct. 27, 2005).

52 2005 Energy Policy Act § 315.

53 Prohibition of Energy Market Manipulation, 70 Fed. Reg. 61,930-61,933 (Oct. 27, 2005) (to be codified at 18 C.F.R. § 159.1).

54 Prohibition of Energy Market Manipulation, 70 Fed. Reg. 61,930-61,933 (Oct. 27, 2005) (to be codified at 18 C.F.R. § 159.1).

55 Prohibition of Energy Market Manipulation, 70 Fed. Reg. 61,930, 61,931 (Oct. 27, 2005) (to be codified at 18 C.F.R. § 159.1).

56 Prohibition of Energy Market Manipulation, 70 Fed. Reg. 61,930, 61,931 (Oct. 27, 2005).

the FERC stated in its market manipulation NOPR that it will require non-jurisdictional entities to adhere to new Section 4A of the NGA.[57]

Thirty parties filed comments on the FERC's NOPR, and on January 19, 2006, the FERC issued a final rule making official the market manipulation regulations that it had proposed in the NOPR.[58] The final rule upheld the FERC's broad reading of "any entity" in determining the scope of the statute's prohibition on market manipulation. The FERC stated:

> If any entity engages in manipulation and the conduct is found to be "in connection with" a jurisdictional transaction, the entity is subject to the Commission's anti-manipulation authority. Absent such nexus to a jurisdictional transaction, however, fraud and manipulation in a non-jurisdictional transaction (such as a first or retail sale) is not subject to the new regulations.[59]

The final rule made no changes to the NOPR's three-part description of the scope of the market manipulation prohibition indicated above. However, the FERC did include a statement of guidance as to how it will apply the new regulation, based on the approach employed by the Securities and Exchange Commission in enforcement actions under its Rule 10b-5.[60] Thus, the FERC stated:

> The Commission will act in cases where an entity: (1) uses a fraudulent device, scheme or artifice, or makes a material misrepresentation or a material omission as to which there is a duty to speak under a Commission-filed tariff, Commission order, rule or regulation, or engages in any act, practice, or course of business that operates or would operate as a fraud or deceit upon any entity; (2) with the requisite scienter; (3) in connection

---

[57] On November 21, 2005, FERC proposed to repeal certain market behavior rules codified at Section 284.288 and 284.403. FERC proposed to repeal the rules because sellers would instead be subject to the more comprehensive market manipulation provisions of the 2005 Energy Policy Act and the new regulations the FERC will issue thereunder. *See* Amendments to Codes of Conduct for Unbundled Sales Service and for Persons Holding Blanket Marketing Certificates, Docket No. RM06-5-000 (Nov. 21, 2005).

[58] *Prohibition of Energy Market Manipulation*, Order No. 670, 114 FERC ¶ 61,047 (2006), 71 Fed. Reg. 4244 (Jan. 26, 2006).

[59] *Prohibition of Energy Market Manipulation*, Order No. 670, 114 FERC ¶ 61,047, at P 16 (2006), 71 Fed. Reg. 4244 (Jan. 26, 2006).

[60] *Prohibition of Energy Market Manipulation*, Order No. 670, 114 FERC ¶ 61,047, at PP 48–49 (2006), 71 Fed. Reg. 4244 (Jan. 26, 2006).

with the purchase or sale of natural gas or electric energy or transportation of natural gas or transmission of electric energy subject to the jurisdiction of the Commission.[61]

[ii] Securities and Commodity Law Precedent

Section 315 of the Act closely tracks the prohibited conduct language in Section 10(b) of the Securities Exchange Act of 1934 (Exchange Act),[62] and specifically dictates that the terms "manipulative and deceptive device or contrivance" are to be used "as those terms are used in Section 10(b)."[63] The Securities and Exchange Commission (SEC) has adopted Rule 10b-5,[64] which implemented Section 10(b) of the Exchange Act, and has developed a significant body of legal precedent related to both Section 10(b) of the Exchange Act and Rule 10b-5. The FERC, consistent with the mandate that its new authority be exercised in a manner consistent with Section 10(b) of the Exchange Act, has modeled its proposed regulations on Rule 10b-5.[65] This approach provides entities subject to the new rule a substantial body of precedent applying the comparable language of Rule 10b-5. According to the FERC, "the vast body of section 10(b) case law provides substantial certainty to entities subject to section 10(b) because it has resolved many recurring questions of interpretation that have arisen under the rule."[66]

Under its market manipulation NOPR, the FERC recognized that Section 4b of the Commodity Exchange Act (CEA) is also intended to provide some guidance for entities subject to the new regulation. Section 4b of the CEA is the Commodity Futures Trading Commission's (CFTC) general anti-fraud rule.[67] Section 4b makes it unlawful

[61] *Prohibition of Energy Market Manipulation*, Order No. 670, 114 FERC ¶ 61,047, at P 49 (2006), 71 Fed. Reg. 4244 (Jan. 26, 2006).

[62] 15 U.S.C. § 78j(b).

[63] 2005 Energy Policy Act § 315.

[64] 17 C.F.R. § 240.10b-5.

[65] One notable exception is that the FERC's proposed new regulation will not create a private right of action. Prohibition of Energy Market Manipulation, 70 Fed. Reg. 61,930, 61,933 (Oct. 27, 2005) (to be codified at 18 C.F.R. § 159.1(b)).

[66] Prohibition of Energy Market Manipulation, 70 Fed. Reg. 61,930, 61,931-61,932 (Oct. 27, 2005).

[67] 7 U.S.C. § 6b.

for any person to "cheat or defraud or attempt to cheat or defraud," or to make false reports or statements, or to deceive or attempt to deceive another in transactions under the CFTC's jurisdiction. Although the statutory language of Section 10(b) of the Exchange Act and Section 4b of the CEA are not identical, the Supreme Court stated that they are "analogous" and "[t]he language of Sec. 4b [of the CEA] is similar to that of Sec. 10(b) of the Securities Exchange Act of 1934."[68] Lower courts have gone further, noting that "the elements of a claim under Sec. 4(b)(A) are basically the same as those under Rule 10b-5."[69] As indicated above, the FERC, in its market manipulation final rule issued on January 19, 2006, provided a statement of guidance on the rule's application that is based on this SEC approach under SEC Rule 10b-5 (*see* § 1.03[3][j][i]).[70]

The January 19, 2006 market manipulation rule also clarifies that the rule does not involve any new affirmative duty to disclose:

> [T]he Final Rule, like section 10(b) of the Exchange Act and SEC Rule 10b-5, is an antifraud provision, not a disclosure provision. Nothing in the Final Rule requires disclosure of sensitive information that would only function to weaken an entity's bargaining position in arm's-length, bilateral negotiations. Absent a tariff requirement or Commission directive mandating disclosure, there is no violation of the Final Rule simply because an entity chooses not to disclose all non-public information in its possession.[71]

Finally, indicating its strong concern that those that manipulate the market are curtailed, Congress amended the NGA to authorize the FERC to enlist the power of the judiciary to prevent a person that has violated the anti-manipulation provisions (new Section 4A) of the NGA from holding a position as an officer or director in a natural

---

[68] Merrill Lynch, Pierce, Fenner & Smith v. Curran, 456 U.S. 353, 389 (1982).

[69] Trustman v. Merrill Lynch, Pierce, Fenner & Smith, 1985 U.S. Dist. LEXIS 23154, at *38; Fed. Sec. L. Rep. (CCH) P91,936 (C.D.C.A. 1985); *see also* Drexel Burnham Lambert, Inc. v. Commodity Futures Trading Comm'n, 850 F.2d 742, 748 (D.C. Cir. 1988) (court held recklessness satisfied the scienter requirement for Section 4b just as it does under Section 10(b) and Rule 10b-5).

[70] *Prohibition of Energy Market Manipulation*, Order No. 670, 114 FERC ¶ 61,047, at PP 48–49 (2006), 71 Fed. Reg. 4244 (Jan. 26, 2006).

[71] *Prohibition of Energy Market Manipulation*, Order No. 670, 114 FERC ¶ 61,047, at P 36 (2006), 71 Fed. Reg. 4244 (Jan. 26, 2006) (internal citations omitted).

gas company or from buying or selling natural gas or transmission services for as long as the court sees fit.[72]

[k] Natural Gas Market Transparency Rules

Section 316 of the Act amends the NGA to require the FERC "to facilitate price transparency in the markets for the sale or transportation of physical natural gas in interstate commerce, having due regard for the public interest, the integrity of those markets, fair competition, and the protection of consumers."[73] Section 316 contains two general provisions to aid in this objective: (1) authorization for the FERC to prescribe rules for the dissemination of market information regarding the availability and price of natural gas, and (2) a directive for the FERC and the CFTC to execute a Memorandum of Understanding with regard to information sharing which addresses, among other things, requests for information and the treatment of proprietary trading information.

[i] FERC Rules for Dissemination of Market Information

The Act permits the FERC to prescribe rules providing for the "timely" dissemination of information about the availability and prices of natural gas sold at wholesale and in interstate commerce to the FERC, state commissions, buyers and sellers of wholesale gas, and to the public.[74] The FERC can obtain this information from any market participant, as well as rely on other entities to receive and make public such information, subject to certain disclosure rules provided in the Act. In promulgating the rules, the FERC must consider the degree of transparency provided by existing price publishers and trade processing services, and must rely on such publishers and services "to the maximum extent possible."[75]

---

[72] 2005 Energy Policy Act § 318 (amending by adding § 20(d) to the Natural Gas Act, 15 U.S.C. § 717 *et seq.*).

[73] 2005 Energy Policy Act § 316 (adding § 23(a)(1) to the Natural Gas Act, 15 U.S.C. § 717 *et seq.*).

[74] 2005 Energy Policy Act § 316 (adding § 23(a)(2) to the Natural Gas Act, 15 U.S.C. § 717 *et seq.*).

[75] 2005 Energy Policy Act § 316 (adding § 23(a)(4) to the Natural Gas Act, 15 U.S.C. § 717 *et seq.*).

The FERC must exempt from disclosure information that the FERC determines would be "detrimental to the operation of an effective market" or would "jeopardize system security" if disclosed. In determining the timing and nature of the information to be made available, the FERC is to seek to ensure that consumers and the market are protected from collusion or other anticompetitive behaviors that could be facilitated by untimely public disclosure of transaction-specific information.[76]

[ii] FERC/CFTC Memorandum of Understanding

Section 316 of the Act requires FERC to execute a Memorandum of Understanding (MOU) with the CFTC to ensure the coordination of information requests to the relevant markets in order to minimize duplicative information requests and to address the treatment of proprietary trading information.

On October 12, 2005, the FERC and the CFTC executed the required MOU.[77] The FERC has agreed to direct to the CFTC the FERC's requests for futures and options trading data or other information to be obtained from a designated contract market, registered derivatives transaction execution facility, or any other board of trade, exchange, or market. The CFTC will obtain the data and furnish the information to the FERC pursuant to the MOU. The data are to be kept confidential and non-public, except (1) in an action to which the FERC, CFTC or United States is a party under Section 8(e) of the CEA, or (2) in certain FERC administrative or enforcement proceedings enumerated in the MOU.

The FERC will also provide information to the CFTC pursuant to the MOU. Any data or information obtained by the FERC pursuant to its confidentiality regulations and furnished under the terms of the MOU is to be kept confidential and non-public and is not to be disclosed by the CFTC except in an action or proceeding to which the CFTC or United States is a party.

---

[76] 2005 Energy Policy Act § 316 (adding § 23(b) to the Natural Gas Act, 15 U.S.C. § 717 *et seq.*).

[77] Memorandum of Understanding Between the Federal Energy Regulatory Commission (FERC) and the Commodity Futures Trading Commission (CFTC) Regarding Information Sharing and Treatment of Proprietary Trading and Other Information (Oct. 12, 2005), *available at* www.ferc.gov and www.cftc.gov.

Under the MOU, the FERC and the CFTC also have agreed:

- to confer on coordinating procedures as to the disclosure of information to conform to third-party notice requirements;
- to avoid duplicative information requests "to the extent possible," and to coordinate on a regular basis on oversight, investigative, and enforcement activities. The respective FERC and CFTC oversight and enforcement staffs are authorized to share information as to these activities to determine mutual interests;
- that aggregate data furnished or derived from data furnished pursuant to the MOU may be disclosed by the agencies and are not subject to the confidentiality provisions of the MOU;
- that neither the FERC nor the CFTC will disclose certain information without the prior written consent of the other agency; and
- that the FERC and the CFTC will notify each other of requests for disclosure of certain information.

[iii] Additional Provisions

Section 316 of the Act provides some important caveats to the new natural gas market transparency provisions. One caveat is that the FERC is not to condition access to interstate pipeline transportation on the reporting requirements of the new section. The Act also prohibits the FERC from requiring natural gas producers, processors, or users who have a "de minimis" market presence to comply with the reporting requirements.

Also, in general, persons are not subject to civil penalties as to any violation occurring more than three years before the date on which the person is provided notice of the proposed penalties under Section 22(b) of the new Act. This provision, however, does not apply if the FERC finds that a seller subject to its jurisdiction has engaged in fraudulent market manipulation activities materially affecting its contract for the transportation or sale of natural gas.

[l] Enhanced Civil and Criminal Penalties

Prior to the passage of the Act, the FERC had a limited arsenal of weapons at its disposal to combat violations of the NGA and the

NGPA.[78] At that time the FERC was authorized to order disgorgement of unjust profits, impose nominal fines, and condition, suspend or revoke market-based rate authority, certificate authority, or blanket certificate authority.[79] However, the Western energy crisis of 2000-2001 highlighted the need for heightened federal penalties with regard to the natural gas industry to deter and punish abusive behavior. Congress subsequently significantly enhanced the FERC's civil and criminal enforcement tools.

[i] Enhanced Criminal Penalties

The Act dramatically increased the FERC's criminal penalty authority under both the NGA and the NGPA. Specifically, Section 314 of the Act amends Section 21 of the NGA and Section 504(c) of the NGPA by increasing the criminal penalties under of each statute.[80] Under these new provisions, violators of either statute are now subject to fines of up to $1 million (increased from $5,000) and imprisonment for up to five years (increased from two years).[81] Any person who willfully and knowingly violates any rule, regulation, restriction, condition, or order made or imposed by the FERC pursuant to either the NGA or the NGPA is also subject to a fine not to exceed $50,000 for each day during which such offense occurs (formerly $500 per day).[82]

[ii] Enhanced Civil Penalties

In addition to the stiffer criminal penalties implemented by the Act, Congress also enhanced the FERC's civil enforcement authority. As Chairman Joseph T. Kelliher noted, this enhancement of the FERC's civil penalty authority was critical to the agency's ability to meet one of the key purposes behind the Act: "Putting FERC's civil penalty muscle on par with those of other federal agencies should be a significant deterrent to any repeat of the sort of unscrupulous behavior

---

[78] Natural Gas Act, 15 U.S.C. § 717 *et seq.*; Natural Gas Policy Act of 1978, 15 U.S.C. § 3301 *et seq.*

[79] *Enforcement of Statutes, Orders, Rules, and Regulations*, 113 FERC ¶ 61,068, at P 4 (2005). Although the FERC is unable to prosecute criminal violations on its own, the FERC is able to refer such matters to the Department of Justice.

[80] 2005 Energy Policy Act § 314.

[81] 2005 Energy Policy Act § 314.

[82] 2005 Energy Policy Act § 314.

that occurred during the Western energy crisis in 2000 and 2001."[83] The FERC's more expansive civil penalty authority operates in tandem with its existing enforcement authority which requires, among other things, the disgorgement of unjust profits.[84]

One key provision of the Act expands the FERC's enforcement power under the NGA. Pursuant to the new Section 22 of the NGA, the FERC is now authorized to assess civil penalties of up to $1 million per day per violation for violations of the NGA or any rules, regulations, restrictions, conditions, or orders of the FERC issued pursuant to the NGA after an opportunity for a public hearing.[85] Prior to the passage of the Act, the FERC was only able to refer NGA violations to the Department of Justice for criminal prosecution. Finally, under Section 314(b)(2) of the Act, any person who knowingly violates any provision of the NGPA, or any provision of any rule or order promulgated pursuant to the NGPA is subject to a civil fine of up to $1 million per violation.

[iii] FERC Policy Statement

To provide "firm but fair enforcement," the FERC issued a Policy Statement on Enforcement on October 20, 2005 (Policy Statement) outlining some of the factors it will consider when developing remedies for violations, including applying the enhanced civil penalty authority provided by the Act. The FERC will review each matter on a case-by-case basis and fashion remedies appropriate to the facts presented.[86]

When determining the appropriate penalty level, the FERC will first look to the "seriousness of the violation."[87] The FERC will consider

---

83 *Energy Bill Provisions Represent Most Significant Changes in Energy Law in 70 Years, FERC Chairman Says*, U.S. Federal News, Aug. 8, 2005, available at 2005 WLNR 12501172.

84 *Enforcement of Statutes, Orders, Rules, and Regulations*, 113 FERC ¶ 61,068, at PP 11–12 (2005).

85 2005 Energy Policy Act § 314(b)(1).

86 *Enforcement of Statutes, Orders, Rules, and Regulations*, 113 FERC ¶ 61,068 (2005).

87 *Enforcement of Statutes, Orders, Rules, and Regulations*, 113 FERC ¶ 61,068, at P 20 (2005). The new Section 22 of the NGA implemented by the Act explicitly requires this. ("In determining the amount of a proposed penalty, the Commission shall take into consideration the nature and seriousness of the violation and the efforts to remedy the violation.").

(1) the extent of the harm resulting from the violation, (2) whether the violation included fraud or deceit, and (3) the violating parties' intent.[88] Second, the FERC will consider efforts the company made to remedy the violation in a timely manner. The FERC may give a company credit based upon the company's internal compliance programs, self-reporting, and the company's exemplary cooperation with the FERC's investigation.[89] In deciding whether to give a company credit for its commitment to compliance, the FERC will consider, among other things, whether the company has a formal, documented, independently-run, and sufficiently-resourced compliance program, and the extent of the entity's compliance training.[90] With regard to self-reporting, the FERC will consider how the market participant uncovered the misconduct, how quickly it notified the FERC, whether any remedial steps were taken, and the extent of the entity's meetings with and disclosures to the FERC. Notably, any such credit would only impact the level of the penalty imposed, not the disgorgement of profits.[91] These factors are consistent with the FERC's past practice and with the practice of other similarly situated federal agencies.

[4] Production

Subtitle C provides several definitional changes to related Acts including the Outer Continental Shelf Lands Act, the Deepwater Port Act, the Safe Drinking Water Act, and the Federal Water Pollution Control Act.[92]

---

[88] *Enforcement of Statutes, Orders, Rules, and Regulations*, 113 FERC ¶ 61,068, at P 20 (2005).

[89] *Enforcement of Statutes, Orders, Rules, and Regulations*, 113 FERC ¶ 61,068, at PP 21–27 (2005).

[90] *Enforcement of Statutes, Orders, Rules, and Regulations*, 113 FERC ¶ 61,068, at PP 21–27 (2005). It is worth noting that, without specifically ordering companies to implement a FERC compliance program, the FERC nonetheless contemplates that companies will have such programs in place.

[91] *Enforcement of Statutes, Orders, Rules, and Regulations*, 113 FERC ¶ 61,068, at P 23 (2005).

[92] 2005 Energy Policy Act §§ 321–323.

[5] Naval Petroleum Reserve

[a] Transfer of Jurisdiction and Control

Section 331 transfers administrative jurisdiction and control over all public domain lands included within Naval Petroleum Reserve Numbered 2 located in Kern County, California[93] from the Secretary of Energy to the Secretary of the Interior.[94] The Act states that the principal purpose for the transfer is to provide for the ultimate economic recovery of the hydrocarbon resources while preventing unnecessary degradation.[95] Furthermore, the Act allows the Secretary of the Interior to dispose of the lands and to allow for commercial or non-profit surface use, not to exceed 10 acres each, provided that such use does not materially interfere with the ultimate economic recovery of the hydrocarbon resources.[96]

[b] Funding Pre-Transfer Environmental Conditions

Section 332 establishes a lease revenue account to fund any and all costs associated with environmental conditions that existed prior to the administrative jurisdiction and control transfer.[97] In addition, to the extent the Secretary of the Interior incurs transition costs associated with the transfer and leasing of the lands, lease revenue account funds will be made available.[98]

---

[93] Located north of Los Angeles and Ventura Counties.

[94] 2005 Energy Policy Act § 331(a). The Act excludes two portions of the Naval Petroleum Reserve Numbered 2 from the transfer: (1) a portion authorized for disposal under Section 3403(a) of the Strom Thurmond National Defense Authorization Act for Fiscal Year 1999 and (2) a 220-acre surface estate conveyed to the City of Taft, California. 2005 Energy Policy Act § 331(b). The federal government reserved all mineral rights to such land, without surface use or occupancy rights. 2005 Energy Policy Act § 333(d).

[95] 2005 Energy Policy Act § 331(c).

[96] 2005 Energy Policy Act § 331(d).

[97] 2005 Energy Policy Act § 332(a)–(b). The lease revenue account must fund all costs and expenses related to investigations, remediation, compliance actions, response, waste management, impediments, fines or penalties related to surface or subsurface conditions that existed prior to the land transfer, as well as any future remediation necessary as a result of any pre-transfer activities. 2005 Energy Policy Act § 331(b).

[98] 2005 Energy Policy Act § 331(b)(2).

The Act provides for funding of the lease revenue account from a pre-transfer source and a post-transfer source. For three years after the transfer of the lands, the lease revenue account will receive $500,000 per year from bonuses, rents, royalties, and interest charges associated with leases entered into before the transfer date.[99] In addition, revenues derived from leases entered on or after the transfer date will be deposited into the lease revenue account.[100] The Act prohibits the lease revenue account from exceeding $3 million.

The lease revenue account terminates upon the Secretary of the Interior certifying (1) that all pre-transfer environmental conditions on the Naval Petroleum Reserve Numbered 2 lands have been successfully completed, (2) that all costs associated with the environmental contamination have been paid in full, and (3) that all transition costs have been paid in full. Any funds remaining, at the time the Secretary of the Interior certifies the termination of the lease revenue account, will be distributed in accordance with the Mineral Leasing Act.[101]

[6] Production Incentives

Congress addressed production incentives in the Act primarily with a broad three-prong approach. Subtitle E of the Act (1) allows for the continuation of the royalty in-kind program; (2) provides a multitude of royalty reduction, suspension, and/or termination programs to encourage development and production activities; and (3) sets forth several scientific data collection, inventorying, sharing, and/or preservation programs. This subtitle also provides several financial assistance programs and addresses orphaned and abandoned wells.

[a] The Royalty In-Kind Program

Section 342 sets forth the policies and administrative procedures that the federal government must follow to request and accept royalty payments in-kind rather than in-value. To request and receive royalty

[99] 2005 Energy Policy Act § 331(c).

[100] 2005 Energy Policy Act § 331(c).

[101] The Mineral Leasing Act provides that "all moneys which may accrue to the United States under the provisions of this chapter and the Geothermal Steam Act of 1970 from lands within the naval petroleum reserves shall be deposited in the Treasury as 'miscellaneous receipts.' " Mineral Leasing Act, 30 U.S.C. § 191. Otherwise, the funds are paid, in part, to the state treasuries and, in part, to the reclamation fund. Mineral Leasing Act, 30 U.S.C. § 191.

payments in-kind, the Secretary of the Interior must determine that "receiving royalties in-kind provides benefits to the United States that are greater than or equal to the benefits that are likely to have been received had royalties been taken in-value."[102]

The Act requires the Secretary of the Interior to submit an annual report describing the methodologies used to conclude that in-kind royalties provide the government greater or equal benefits to in-value royalties, as well as the supporting cost/benefit analysis.[103] The Act prohibits any revenues from the sale of in-kind royalties to pay for federal government personnel, travel, or other administrative costs, except for those salaries and other administrative costs directly related to the royalty in-kind program.[104]

The Act also requires the Secretary of the Interior to consult with the relevant state government prior to conducting a royalty in-kind program, and to the extent allowed by federal law, the Secretary may delegate management of the program, or any portion thereof, to that particular state government.[105] In addition, the Act requires the Secretary to consult with each state where the royalty in-kind program has been instituted "to ensure, to the maximum extent practicable" that the in-kind program is achieving benefits greater than or equal to those that would be received if royalties were taken in-value.[106]

The Act also establishes two preference provisions within the royalty in-kind program: one for small refineries and one for low-income assistance. First, the Act authorizes the Secretary of the Interior to grant a preference to certain refineries for purchasing federal royalty

---

102 2005 Energy Policy Act § 342(d).

103 2005 Energy Policy Act § 342(e). The Act requires the Secretary of the Interior to report the following four categories of information: (1) the methodologies used to determine in-kind benefits would be greater than or equal to in-value benefits, including the performance standard for comparing in-kind and in-value amounts; (2) an explanation describing the evaluation performed to make the determination that the in-kind benefits would b greater than or equal to the in-value benefits; (3) the actual amounts received from taking the royalties in-kind and the costs/saving incurred; and (4) an evaluation of other relevant public benefits or detriments associated with taking royalties in-kind. 2005 Energy Policy Act § 342(e)(2)(A)–(D).

104 2005 Energy Policy Act § 342(b)(4)–(5).

105 2005 Energy Policy Act § 342(g).

106 2005 Energy Policy Act § 342(g).

volumes.[107] To grant such a preference, the Secretary must find that "sufficient supplies of crude oil are not available in the open market to refineries that do not have their own source of supply for crude oil."[108] These refineries may purchase the preference volumes for processing or for use, and such purchases may occur through a private sale.[109] The Act requires the royalty volumes be sold at not less than market price.[110] To the extent more than one refinery does not have its own source of supply and sufficient supplies are not available in the open market, the Act allows the Secretary of the Interior to prorate the preferential volumes among such refineries.[111]

The Act also authorizes the Secretary of the Interior to grant a preference "to any person, including any Federal or State agency, for the purpose of providing additional resources to any Federal low-income energy assistance program."[112] If the Secretary of the Interior grants such a preference, the Act requires the Secretary to submit a report to Congress, within three years, assessing the effectiveness of granting the preference and providing a specific recommendation as to the continuation of authority to grant preferences.[113]

[b] Marginal Properties

Marginal wells contribute significantly to this nation's energy resources. In 2003, 400,000 marginal oil wells produced approximately 15% of the United States' domestic oil production and 7% of the natural gas production.[114] From these wells, in excess of 313 million barrels of oil were produced.[115]

---

[107] 2005 Energy Policy Act § 342(h)(1).

[108] 2005 Energy Policy Act § 342(h)(1).

[109] 2005 Energy Policy Act § 342(h)(1).

[110] 2005 Energy Policy Act § 342(h)(1).

[111] 2005 Energy Policy Act § 342(h)(2).

[112] 2005 Energy Policy Act § 342(j).

[113] 2005 Energy Policy Act § 342(j)(2).

[114] The Energy Policy Act of 2005: Hearings Before the Subcommittee on Energy and Air Quality of the House of Representatives Committee on Energy and Commerce, 109th Cong. 3-5 (2005) (prepared Statement of Victor Carrillo, Chairman, Texas Railroad Commission).

[115] The Energy Policy Act of 2005: Hearings Before the Subcommittee on Energy and Air Quality of the House of Representatives Committee on Energy and Commerce, 109th Cong. 3-5 (2005) (prepared Statement of Victor Carrillo, Chairman, Texas Railroad Commission).

The Act provides for a reduction in royalty rates as an incentive for producers to keep marginal wells online. Congress established the initial standards and requirements for onshore marginal properties in the Act. However, Congress deferred to the Secretary of the Interior's rulemaking authority for Outer Continental Shelf properties and authorized the Secretary of the Interior to prescribe different standards and requirements for onshore marginal property royalty relief.[116]

The Act defines an onshore marginal property as one that produces, on average, less than 15 barrels of oil equivalent per day or 90 MMBtu of gas equivalent per day.[117] For the Secretary of the Interior to trigger royalty relief, crude must trade[118] under $15/barrel for 90 consecutive days and natural gas must trade[119] under $2.00/MMBtu for 90 consecutive days.[120] Once triggered, the royalty rate becomes the lesser of (a) 5% or (b) the statutory or regulatory rate applicable to the affected production.[121] The royalty rate reduction incentive terminates if the property no longer meets the definition of "marginal" or the spot prices exceed the thresholds mentioned above for 90 consecutive trading days.[122]

The Act requires the Secretary of the Interior, within 18 months of enactment of the Act, to develop the standards and requirements for marginal properties located in the Outer Continental Shelf.[123] The Secretary must consider the following factors when developing the standards for marginal properties: (1) oil and gas prices and market trends, (2) production costs, (3) abandonment costs, (4) federal and state tax provisions and the effects of those provisions on production

[116] 2005 Energy Policy Act § 343(e)(1).

[117] 2005 Energy Policy Act § 343(a). The most recent three months of production are used for calculating the average daily production rates. 2005 Energy Policy Act § 343(a).

[118] Defined as the West Texas Intermediate (WTI) spot price at Cushing, Oklahoma.

[119] Defined as the natural gas price at Henry Hub, Louisiana.

[120] 2005 Energy Policy Act § 343(b).

[121] 2005 Energy Policy Act § 343(c).

[122] 2005 Energy Policy Act § 343(d).

[123] 2005 Energy Policy Act § 343(e)(2). If the Secretary of the Interior determines that it is not practicable to issue such regulations, the Secretary must provide a report to Congress explaining that determination. 2005 Energy Policy Act § 343(e)(2).

economics, (5) other royalty relief programs, (6) regional differences in average wellhead prices, (7) national energy security issues, and (8) other relevant matters.[124]

[c] Gulf of Mexico's Shallow and Deep Waters

The Act provides royalty relief for both shallow and deep water wells. The Act requires the Secretary of the Interior to issue regulations granting royalty relief[125] to "ultra deep well" leases issued in the Gulf of Mexico shallow waters.[126] Congress defined shallow waters as those waters less than 400 meters deep (approximately 1300 feet)[127] and an "ultra deep well" as "a well drilled with a perforated interval, the top of which is at least 20,000 true vertical depth below the datum at mean sea level."[128] Despite the Act's use of meters to measure the shallow water's depth, presumably, the 20,000 refers to feet.

The Act also provides royalty relief for deep wells on leases issued in waters more than 200 meters deep but less than 400 meters deep, and for wells located in water depths greater than 400 meters, all located within the same area of the Gulf of Mexico as the ultra deep wells in shallow waters.[129]

[d] Oil and Gas Leasing in the National Petroleum Reserve in Alaska

Section 347 of the Act revises and adds significant provisions to the Naval Petroleum Reserves Production Act of 1976.[130] The new section provides that each lease in the National Petroleum Reserve in Alaska will be for an initial lease term of 10 years, with automatic extension as long thereafter as oil or gas is produced (or capable of being produced) in paying quantities. Leases can be renewed as long as the lessee certifies that hydrocarbon resources have been discovered in such quantities that a prudent operator would hold the land for

---

[124] 2005 Energy Policy Act § 343(e)(4)(A)–(H).

[125] Suspension volumes must not be less than 35 Bcf. 2005 Energy Policy Act § 344(a)(2).

[126] 2005 Energy Policy Act § 344(a)(1).

[127] 2005 Energy Policy Act § 344(a)(1).

[128] 2005 Energy Policy Act § 344(a)(3).

[129] 2005 Energy Policy Act §§ 344(b), 345.

[130] 42 U.S.C. § 6501 *et seq.*

further development.[131] The lessee also may renew its lease even if no such discovery has occurred if the lessee pays $100/acre renewal fee and provides evidence that diligent exploration efforts have been made that warrant continuation and that continuation is intended.[132] If no gas or oil is produced from the lease within 30 years after the issuance of the lease, the lease expires.[133] The Act provides a force-majeure-of-sorts provision in that no lease issued under Section 347 will expire if the lessee failed to produce for circumstances beyond its control.[134]

[e] Unitization Agreements

The Act also provides for unitization of a pool, field, reservoir, or like area if the Secretary of the Interior determines unitization is in the public interest.[135] In making this determination, the Secretary must consider, among other things, the extent to which the unit agreements will minimize the impact to surface resources of the leases and will facilitate consolidation of facilities.[136] The Secretary must consult with the State of Alaska in making the determination.[137] The Secretary of the Interior is required to utilize a production allocation methodology for any unit that includes land other than federal land in the Reserve.[138]

[f] Exploration Incentives

With respect to leases that take effect after the enactment of the Act, the Secretary of the Interior can provide exploration incentives to encourage the greatest ultimate recovery of oil and gas.[139] If, in the Secretary of the Interior's judgment, the Secretary believes it necessary to promote development or where a lease cannot be successfully operated under its terms, the Secretary may waive, suspend, or

[131] 2005 Energy Policy Act § 347 (amending 42 U.S.C. § 6508).

[132] 2005 Energy Policy Act § 347 (amending 42 U.S.C. § 6508).

[133] 2005 Energy Policy Act § 347 (amending 42 U.S.C. § 6508).

[134] 2005 Energy Policy Act § 347 (amending 42 U.S.C. § 6508).

[135] 2005 Energy Policy Act § 347 (amending 42 U.S.C. § 6508).

[136] 2005 Energy Policy Act § 347 (amending 42 U.S.C. § 6508).

[137] 2005 Energy Policy Act § 347 (amending 42 U.S.C. § 6508).

[138] 2005 Energy Policy Act § 347 (amending 42 U.S.C. § 6508).

[139] 2005 Energy Policy Act § 347 (amending 42 U.S.C. § 6508).

reduce rental fees or royalties.[140] The Secretary of the Interior is required to consult with the State of Alaska and the North Slope Borough of Alaska and any applicable Regional Corporation.[141]

[g] North Slope Science Initiative

Congress took a step towards further development of the Alaskan North Slope by establishing the North Slope Science Initiative (Initiative). The Initiative's purpose is to coordinate the collection of scientific data to better understand the terrestrial, aquatic, and marine ecosystems of the Alaskan North Slope.[142] The minimum objectives focus on information needs and access to information regarding past and anticipated development of the North Slope, natural resource management, funding, scientific approach, and peer review.[143]

The Act provides for a panel of not more than 15 scientists and technical experts from a diverse group, including the oil and gas industry, subsistence users, Native Alaskan entities, conservation organizations, wildlife management organizations, and academia.[144] Within three years, and then every year thereafter, the Secretary of the Interior must publish a report describing the Initiative's studies and findings.[145] The Act provides "such sums" of funds "necessary to carry out" the Initiative.[146]

---

140 2005 Energy Policy Act § 347 (amending 42 U.S.C. § 6508).

141 2005 Energy Policy Act § 347 (amending 42 U.S.C. § 6508).

142 2005 Energy Policy Act § 348(a).

143 2005 Energy Policy Act § 348(b). The specific minimum objectives include: (1) identify and prioritize information needs for inventory, monitoring, and research activities; (2) develop an understanding of information needs; (3) focus on prioritization of natural resource management and ecosystem information; (4) coordinate ongoing and future inventory, monitoring, and research activities to minimize duplication of effort, and share financial resources and expertise; (5) identify priority needs and develop a funding strategy; (6) provide a consistent approach to high caliber science; (7) maintain and improve access to research and knowledge; and (8) ensure through peer review that science is of the highest technical quality. 2005 Energy Policy Act § 348(c)(1)–(8).

144 2005 Energy Policy Act § 348(c).

145 2005 Energy Policy Act § 348(e).

146 2005 Energy Policy Act § 348(f).

[h] Orphaned, Abandoned, and Idled Wells

The number of orphaned wells in the United States has been estimated at 57,000.[147] As the name suggests, these are wells that no longer produce or are idle without approval of the state, and for which the operators are unknown or have become insolvent.[148] Within one year of the enactment of the Act, the Secretary of the Interior, in cooperation with the Secretary of Agriculture, must establish a program to remediate, reclaim and close orphaned, abandoned, or idled oil and gas wells.[149] The program must address cost recovery, financial assurances, and a means to priority rank the wells. During that same year, the Secretary of the Interior, also in cooperation with the Secretary of Agriculture, must submit to Congress a plan for carrying out the program.[150]

The Act requires the Secretary of the Interior to provide for an orphaned well reimbursement pilot program. Congress appropriated $5 million annually (of the $25 million annual appropriation for the orphaned, abandoned, and idled well total program addressed in Section 349 of the Act) for the fiscal years 2006-2010 to fund this two-prong reimbursement program.[151]

For new federally-owned land leases, the Secretary of the Interior may require the lessee to remediate, reclaim and close all orphaned wells on the leased land.[152] For these newly leased lands, the Act requires the Secretary of the Interior to develop a program to reimburse the lessee for reasonable actual costs for requiring the lessee to remediate, reclaim and close such orphaned wells.[153] The

---

[147] The Energy Policy Act of 2005: Hearings Before the Subcommittee on Energy and Air Quality of the House of Representatives Committee on Energy and Commerce, 109th Cong. 3-5 (2005) (prepared Statement of Victor Carrillo, Chairman, Texas Railroad Commission).

[148] The Energy Policy Act of 2005: Hearings Before the Subcommittee on Energy and Air Quality of the House of Representatives Committee on Energy and Commerce, 109th Cong. 3-5 (2005) (prepared Statement of Victor Carrillo, Chairman, Texas Railroad Commission).

[149] 2005 Energy Policy Act § 349(a).

[150] 2005 Energy Policy Act § 349(d).

[151] 2005 Energy Policy Act § 349(h).

[152] 2005 Energy Policy Act § 349(f)(1).

[153] 2005 Energy Policy Act § 349(f)(1).

reimbursement program will operate through a royalty credit or "other means."[154] For unleased federal lands or for existing federally-owned land leases, the Secretary of the Interior may authorize the lessee to reclaim an orphaned well.[155] Interestingly, the reimbursement provision relating to other federal lands or existing leases expressly calls for 100% reimbursement of reasonable actual costs of remediation, reclamation, and closing of the orphaned well, as opposed to the new lease provision which merely allows for reimbursement of reasonable actual costs.[156]

Finally, the Act also requires the Secretary of Energy to establish a program for providing technical and financial assistance to oil and gas producing states to ensure a practical and economical remedy for addressing environmental problems caused by orphaned or abandoned well sites.[157] The program must cover well sites located on state or private lands. The program should include mechanisms to facilitate the identification of financial assurances, ranking criteria, remediation, best practices information and training, and funding of the state's mitigation efforts on a cost-shared basis.[158] Notwithstanding the $5 million annual appropriation for the reimbursement pilot program discussed above, Congress appropriated $25 million annually to address orphaned, abandoned, and idled wells.[159]

[i] Combined Hydrocarbon Leasing

The Act amends Section 17(b)(2) of the Mineral Leasing Act[160] to allow the Secretary of the Interior to separately lease any area that contains both tar sand and oil and/or gas.[161] The bidding process, annual rental, and posting period for leases issued for tar sand will be the same as that for oil and gas issued leases, except that the minimum acceptable bid for a tar sand lease will be $2/acre.[162]

[154] 2005 Energy Policy Act § 349(f)(1).

[155] 2005 Energy Policy Act § 349(f)(2).

[156] *Compare* 2005 Energy Policy Act § 349(f)(1)(B), *with* 2005 Energy Policy Act § 349(f)(2)(B).

[157] 2005 Energy Policy Act § 349(g).

[158] 2005 Energy Policy Act § 349(g)(3).

[159] 2005 Energy Policy Act § 349(h).

[160] 30 U.S.C. § 226(b)(2).

[161] 2005 Energy Policy Act § 350(a).

[162] 2005 Energy Policy Act § 350(a).

Importantly, the Act grants the Secretary of the Interior the right to waive, suspend, or alter any requirement that a permittee, prospecting for tar sand, exercise due diligence to promote any resource covered by a combined hydrocarbon lease.[163]

[j] Preservation of Geological and Geophysical Data

In Section 351 of the Act, Congress sought to preserve geological and geophysical data by creating a national catalog of archived geological, geophysical, and engineering data, maps, well logs, and samples, and a system to archive such data. The Act also provides technical and financial assistance related to the archived materials.[164] Within a year of the enactment of the Act, the Secretary of the Interior must submit a plan to Congress for implementing the preservation program.[165] The national catalog will be accessible to the public, with regard given to requirements related to confidentiality and proprietary data.[166]

The Act requires the Advisory Committee created by the National Geologic Mapping Act of 1992[167] to advise the Secretary with respect to the preservation program.[168] Congress appropriated $30 million annually for the fiscal years 2006-2010 to carry out the preservation program.[169] The Act incorporates several funding limitation provisions. First, the Act expressly limits the federal share of the financial assistance to 50% of the costs of the activity. Second, the Act applies any private contributions to the non-federal share of the costs.[170] Third, the Act requires the Secretary of the Interior to provide financial assistance for archival facilities and for studies, technical understanding and interpretation assistance, and use of archived materials only to the extent appropriated funds are available.[171] Moreover, Congress

163 2005 Energy Policy Act § 350(a).

164 2005 Energy Policy Act § 351(b).

165 2005 Energy Policy Act § 351(c).

166 2005 Energy Policy Act § 351(e).

167 43 U.S.C. § 31a *et seq.*

168 2005 Energy Policy Act § 351(f).

169 2005 Energy Policy Act § 351(k).

170 2005 Energy Policy Act § 351(g)(3)–(4).

171 2005 Energy Policy Act § 351(g)(1)–(2).

expressly stated that its intent was not for "States [to] use this section as an opportunity to reduce State resources applied to the [preservation] activities."[172]

[k] Oil and Gas Lease Acreage Limitations

Section 27(d)(1) of the Mineral Leasing Act[173] limited any one person, company, or similar entity from holding more than 246,080 acres of oil and gas leases in any one state other than Alaska, except if the "acreage [is] held in special tar sand areas." Section 352 of the Act broadened that exception to also allow a person, company, or similar entity to exceed that limit if the "acreage [is] under any lease any portion of which has been committed to a federally approved unit or cooperative plan or communitization agreement or for which royalty (including compensatory royalty or royalty in-kind) was paid in the preceding calendar year."[174]

[l] Royalty Incentives for the Production of Natural Gas Hydrate Resources

To encourage the production of natural gas from natural gas hydrate resources on the Outer Continental Shelf and the onshore federal lands in Alaska, the Act allows the Secretary of the Interior to conduct a rulemaking and to grant royalty relief for leases issued in the identified regions before January 1, 2016, and under which natural gas production from gas hydrate resources commences prior to January 1, 2018.[175]

The Act requires the Secretary of the Interior to conduct a rulemaking and grant royalty relief if the Secretary determines that such relief would encourage production of natural gas from eligible leases from gas hydrate resources.[176] The maximum suspension volume[177] is 30 Bcf per lease, and the suspension volume applies to any eligible production occurring on or after the date of publication of the advanced

172 2005 Energy Policy Act § 351(i).

173 30 U.S.C. § 184(d)(1).

174 2005 Energy Policy Act § 352.

175 2005 Energy Policy Act § 353(a)–(b).

176 2005 Energy Policy Act § 353(b)(3).

177 Royalty suspension volume means a volume of production from a lease that is not subject to royalty under the provisions of this part. 30 U.S.C. § 203.

notice of proposed rulemaking.[178] The royalty relief is in addition to any other royalty relief that does not specifically grant a gas hydrate production incentive.[179] Limitations, based on market price, may be placed on the royalty relief granted.[180]

The Secretary of the Interior, after consulting the Secretary of Energy, must review other opportunities to enhance production of natural gas from gas hydrate resources on the Outer Continental Shelf and Alaskan federal lands and submit a report to Congress within one year of the enactment of the Act.[181]

[m] Enhanced Production Through Carbon Dioxide Injections

Premised on Congress' findings that approximately two-thirds of the original oil in place in the United States remains unproduced and that use of carbon dioxide has the potential to increase oil and natural gas production, the Act provides royalty incentives for the use of carbon dioxide and other gas injection enhanced recovery techniques.[182] If the Secretary of the Interior determines that the reduction of a royalty for certain leases would be in the public interest and promotes oil and natural gas production, the Secretary shall conduct a rulemaking to provide for such reductions in royalty for the eligible leases.[183]

The Secretary of the Interior's rulemaking authority is limited by volume and price. The suspension volume cannot exceed 5 million barrels of oil equivalent per lease and such volumes must be applied to production occurring on or after the publication of any advance notice of the proposed rulemaking.[184] Additionally, the Secretary of the Interior may place limitations on the amount of royalty reduction based on market price.[185]

178 2005 Energy Policy Act § 353(b)(3).

179 2005 Energy Policy Act § 353(b)(3).

180 2005 Energy Policy Act § 353(b)(4).

181 2005 Energy Policy Act § 353(e).

182 2005 Energy Policy Act § 354(a).

183 2005 Energy Policy Act § 354(b).

184 2005 Energy Policy Act § 354(b)(4).

185 2005 Energy Policy Act § 354(b)(5).

Section 354(c) requires the Secretary of Energy to establish a competitive grant program for oil and gas producers to carry out carbon dioxide injection enhanced recovery projects, while increasing the sequestration of carbon dioxide.[186] The program applies to projects in the Williston Basin and the Cook Inlet Basin. The Act limits the number of projects that can be undertaken and sets forth minimum project requirements that the grant applicants must fulfill.

When evaluating and selecting the grant recipients the Secretary of Energy must consider each applicant's previous experience with gas injection enhanced recovery projects. Moreover, the Secretary of Energy must give priority consideration to applications that "(i) are most likely to maximize production . . . in a cost-effective manner; (ii) sequester significant quantities of carbon dioxide . . .; (iii) demonstrate the greatest commitment . . . to ensure funding for the proposed project and the greatest likelihood that the project will be maintained or expanded . . .; and (iv) minimize any adverse environmental effects from the project."[187]

The Act placed several restrictions on the grant program including: (1) the Secretary of Energy cannot provide more than $3 million in federal assistance to any applicant; (2) the Secretary of Energy must require cost sharing; (3) any project funded by the grant program must begin construction within two years of the date of provision of the grant, but no later than December 31, 2010; (4) the Secretary of Energy cannot provide grant funds to any applicant for more than five years; and (5) the Secretary of Energy must establish procedures to ensure information and knowledge gained by the grant program participants are transferred among other participants and interested persons.[188]

[n] Assessment of Hawaii's Dependence on Oil

Section 355 of the Act charges the Secretary of Energy with assessing the economic implications of Hawaii's dependence on oil as the principal source of energy for the state.[189] Included in its studies, the Secretary of Energy is required to look at both long- and short-term prospects for crude oil supply disruptions as well as the

186 2005 Energy Policy Act § 354(c).

187 2005 Energy Policy Act § 354(c)(4)(A)–(B).

188 2005 Energy Policy Act § 354(c)(5)–(6).

189 2005 Energy Policy Act § 355(b).

associated price volatility and impact on the state's economy.[190] The Act also requires, among other things, an analysis of the technical and economic feasibility (1) of increasing the contribution of renewable energy resources and of using LNG to displace residual fuel oil for generating electricity and (2) of using renewable energy sources for ground, marine, and air transportation energy applications to displace the use of refined petroleum products.[191] The Secretary of Energy, along with state agencies and others, must submit a report to Congress within 300 days of the enactment of the Act, describing their findings, conclusions, and recommendations.[192] The Act provides for the appropriation of funds necessary to carry out the assessment.[193]

[o] The Denali Commission

Congress appropriated up to $55 million annually for fiscal years 2006-2015 in order for the Denali Commission, established in 1998, to carry out energy programs in Alaska.[194] These energy programs include energy generation and development, construction of energy transmission, replacement and cleanup of fuel tanks, construction of fuel transportation networks and related facilities, power cost equalization programs, and projects using coal as a fuel.[195] The Act also required the Denali Commission meetings to be open to the public.[196]

[p] Comprehensive Inventory of Outer Continental Shelf Oil and Natural Gas Resources

In an effort to better understand the Outer Continental Shelf resources, the Act directs the Secretary of the Interior to conduct an inventory and analysis of all Outer Continental Shelf oil and natural gas resources, including offshore Mexico and Canada.[197] Among other things, the analysis must include the change in resource estimates with respect to gathering geological and geophysical data, initial exploration

[190] 2005 Energy Policy Act § 355(b).

[191] 2005 Energy Policy Act § 355(b).

[192] 2005 Energy Policy Act § 355(c).

[193] 2005 Energy Policy Act § 355(d).

[194] 2005 Energy Policy Act § 356.

[195] 2005 Energy Policy Act § 356(b).

[196] 2005 Energy Policy Act § 356(c).

[197] 2005 Energy Policy Act § 357.

and full field development.[198] Furthermore, the Act requires an estimate of the effect that understated inventories have on domestic energy investments. The Act also requires an identification and explanation of how legislative, regulatory, and administrative programs have restricted or impeded the development of these resources, as well as the extent such programs affect domestic supply. The Secretary of the Interior is required to submit a report to Congress on the resource inventories and the analysis of restrictions and impediments within six months of the enactment of the Act, which shall be publicly available, and which shall be updated at least every five years.[199]

[7] Access to Federal Lands

As discussed above, Title III of the 2005 Energy Policy Act implements a wide range of incentives designed to promote adequate growth in the supply of natural gas and oil resources in the United States. Among the incentives set forth under the Act is improved access to federal lands and an overhaul of the permitting and management practices for public land oil and gas leasing. To this end, the Act places requirements on the Secretaries of the Interior and Agriculture to speed up the oil and gas lease application process and establish coordination of the leasing/permitting efforts of both Departments.[200] The Act also establishes the Oil Shale, Tar Sands and Other Strategic Unconventional Fuel Act of 2005 to coordinate and facilitate development of unconventional fuel for research and potential commercial leasing on public lands.[201]

[a] Federal Onshore Oil and Gas Leasing and Permitting Practices

The Act requires the Secretary of the Interior to perform an internal review of current federal onshore oil and gas leasing and permitting practices.[202] The review is to include an examination of the processes for (1) accepting or rejecting offers to lease; (2) administrative appeals

---

198 2005 Energy Policy Act § 357(a).

199 2005 Energy Policy Act § 357(b).

200 2005 Energy Policy Act § 362(b).

201 2005 Energy Policy Act § 369.

202 2005 Energy Policy Act § 361.

of decisions or orders of the Bureau of Land Management regarding federal oil and gas leases; (3) considering surface use plans of operation, including the timeframes in which the plans are considered and any recommendations for improving or expediting the process; and (4) identifying stipulations to address site-specific concerns and conditions, including those stipulations relating to environmental and resource-use conflicts. The Act requires the review to be performed in consultation with the Secretary of Agriculture with regard to any National Forest System lands. The Secretaries of the Interior and Agriculture must submit a joint report to Congress no later than 180 days after enactment of the Act. The report must describe actions taken or plans to improve the federal onshore oil and gas leasing program.[203]

[b] Management of Federal Oil and Gas Leasing Programs

The Act calls for timely action by the Secretaries of the Interior and Agriculture on oil and gas leases and applications for permits to drill on land otherwise for leasing.[204] Timely action includes ensuring "expeditious compliance" with applicable environmental and cultural resources laws,[205] improving the collection, storage and retrieval of information related to leasing activities,[206] and improving inspection and enforcement of oil and gas permitting activities.[207] Appropriations are granted under Title III for these activities.[208]

---

203 2005 Energy Policy Act § 361(b). Under § 361(b) of the Act, the report issued to Congress is to include a description of actions taken under Section 3 of Executive Order No. 13,212, 42 U.S.C. § 13201 (2003). Section 3 of Executive Order 13,212 established an Interagency Task Force within the Department of Energy to, in part, "monitor and assist the agencies in their efforts to expedite their reviews of permits or similar actions, as necessary, to accelerate the completion of energy-related projects (including pipeline safety projects), increase energy production and conservation, and improve the transmission of energy."

204 2005 Energy Policy Act § 362.

205 The Secretary of the Interior must comply with environmental review requirement under NEPA § 102(2)(C), 42 U.S.C. § 4332(2)(C). 2005 Energy Policy Act § 362(a)(1)(A).

206 The Secretary of the Interior must also improve consultation and coordination with the states and the public. 2005 Energy Policy Act § 362(a)(1)(B).

207 2005 Energy Policy Act § 362(c).

208 2005 Energy Policy Act § 362(d).

The Secretary of the Interior must develop best management practices to ensure both timely action on leases and permit applications and improved administration of the onshore oil and gas leasing program under the Mineral Leasing Act.[209] The Secretary of Agriculture is also required to propose regulations to establish specific timeframes for approving, disapproving and appealing resource management plans, lease applications, applications for permits to drill and surface use plans.[210] In developing the best management practices, the Secretary of the Interior must consider any recommendations resulting from the internal review of oil and gas leasing and permitting practices, discussed above (*see* § 1.03[7][a]).

[c] Consultation Regarding Oil and Gas Leasing on Public Land

The Act recognizes the need for better coordination and consistency between public land oil and gas leases administered by the Department of the Interior and National Forest System oil and gas leases administered by the Department of Agriculture. As a result, the Secretaries of the Interior and Agriculture must enter into a Memorandum of Understanding (MOU) to establish coordinated administrative procedures for the timely processing of lease applications, drilling permit applications and surface use plans for oil and gas leases under their respective jurisdictions.[211] The MOU must eliminate duplicative efforts in planning, data management, application tracking and environmental compliance.[212] The MOU also must provide measures for ensuring that lease stipulations are coordinated, applied consistently between the agencies, and "only as restrictive as necessary to protect the resource for which the stipulations are applied."[213]

[d] Estimates of Oil and Gas Resources Underlying Onshore Federal Land

The Energy Act of 2000, among other things, required the Secretary of the Interior, in consultation with the Secretaries of Agriculture and

---

209 30 U.S.C. § 181 *et seq.* 2005 Energy Policy Act § 362(b)(1)(A).

210 2005 Energy Policy Act § 362(b)(3).

211 2005 Energy Policy Act § 363.

212 2005 Energy Policy Act § 363(b).

213 2005 Energy Policy Act § 363(b)(3)(C).

Energy, to conduct an inventory of the estimates of oil and gas reserves of all onshore federal lands and to update it regularly.[214] The Act amends Section 604 of the Energy Act of 2000 to require the Secretaries to focus on "resource" estimates instead of "reserve" estimates, which would appear likely to avoid complications related to whether the focus must be on proven reserves. The Act also requires that the inventory identify restrictions or impediments to the development of oil and gas resources, including delays in granting of leases, drilling permit applications, processing of environmental permits or production and post-lease restrictions.[215] The Act eliminates authorization for appropriations to implement Section 604 of the Energy Act of 2000.[216]

[e] Pilot Project to Improve Federal Permit Coordination

In its effort under the Act to streamline the oil and gas leasing and drilling permit application processes, Congress required the Secretary of the Interior to develop the Federal Permit Streamlining Pilot Project.[217] Under an MOU with the Secretary of Agriculture, the EPA Administrator and the Chief of Engineers,[218] the Secretary of the Interior is to organize a team of personnel with particular expertise in permitting and regulatory matters under the Endangered Species Act,[219] Federal Water Pollution Control Act,[220] Clean Air Act[221] and National Forest Management Act[222] in certain designated Pilot Project field offices.[223] The Pilot Project is to provide applicants with

[214] 42 U.S.C. § 6217(b).

[215] 2005 Energy Policy Act § 364.

[216] *See* 2005 Energy Policy Act § 365(d), amending subsection (d) of the Energy Act of 2000.

[217] 2005 Energy Policy Act § 365.

[218] The Secretary of the Interior may also request that the governors of Wyoming, Montana, Colorado, Utah and New Mexico serve as signatories to the MOU. 2005 Energy Policy Act § 365(b)(2).

[219] 16 U.S.C. § 1536.

[220] 33 U.S.C. § 1344.

[221] 42 U.S.C. § 7401 *et seq.*

[222] 16 U.S.C. § 472a *et seq.*

[223] The Act designates the following BLM Field Offices as Pilot Project offices: Rawlins, Wyoming; Buffalo, Wyoming; Miles City, Montana; Farmington, New Mexico; Carlsbad, New Mexico; Grand Junction/Glenwood Springs, Colorado; and Vernal, Utah. 2005 Energy Policy Act § 365(d).

one-stop-shopping to streamline the application process and to generate coordinated review of proposed energy projects, planning and environmental analyses. In furtherance of that goal, the Act establishes a "BLM Permit Processing Improvement Fund" under Section 35 of the Mineral Leasing Act.[224] Fifty percent of rentals received from leases in any state (excluding Alaska) shall be allocated toward the fund to be appropriated by the Secretary of the Interior for coordinating and processing onshore oil and gas use authorizations on federal land under the jurisdiction of the Pilot Project field offices.[225]

[f] Deadline for Consideration of Applications for Permits

The Act amends the Mineral Leasing Act (MLA)[226] to establish a timeline under which the Secretary of the Interior must consider permit applications under the MLA. Under the new provisions to the MLA, a permit can be issued in as short as 30 days, furthering Congress' intent to lessen the burden on those seeking to develop domestic oil and natural gas supplies. Under the new timeline, the Secretary of the Interior has 10 days after receipt of a permit application to notify the applicant as to whether the application is complete or whether information is missing.[227] The notice must specify the information needed to complete the application. No later than 30 days after submittal of a complete permit application, the Secretary of the Interior must issue the permit if requirements under the NEPA and other environmental laws have been complied with by that time or defer acting on the application.[228] If the permit is deferred, the Secretary must notify the applicant of any steps the applicant could take to be granted the permit, a list of actions the agency must take to comply with applicable laws, and a timeline showing deadlines for completing such actions.[229] Upon deferral, the applicant has two years from the date of the receipt of the deferral notice to complete all the requirements specified.[230] Once the requirements have been met, the

[224] 30 U.S.C. § 191.

[225] 2005 Energy Policy Act § 365(g).

[226] 30 U.S.C. § 226. The additional language was added in § 226(p).

[227] 2005 Energy Policy Act § 366.

[228] 2005 Energy Policy Act § 366.

[229] 2005 Energy Policy Act § 366.

[230] 2005 Energy Policy Act § 366.

Secretary of the Interior has no more than 10 days to issue a decision on the permit application.[231] If the applicant fails to provide the specified information within two years of receipt of the notice, the Secretary must deny the permit.

[g] Fair Market Value Determinations for Linear Rights-of-Way Across Public Lands and National Forests

The Act requires the Secretaries of the Interior and Agriculture to use the same fair market value rent of linear rights-of-way across public lands and National Forests.[232] The fair market value rent will be determined in accordance with 43 C.F.R. Subpart 2806. Within one year of enactment of the Act, the Secretary of the Interior must update 43 C.F.R. Subpart 2806. The Act requires 43 C.F.R. Subpart 2806 to be updated to revise the per acre rental fee zone value schedule by state, county and type of linear right-of-way use to reflect current values of land in each zone.[233]

[h] Energy Right-of-Way Corridors on Federal Land

The Act requires energy right-of-way corridors on federal land to be designated within the 11 contiguous western states, as defined by the Federal Land Policy and Management Act,[234] by the Secretaries of Agriculture, Commerce, Defense, Energy and Interior, in consultation with the FERC, the states, tribal or local units of government, affected utility industries and other interested persons. The corridors shall be designated within two years of enactment of the Act.[235] Energy right-of-way corridors on federal land within other states should be designated within four years of enactment.[236] The Act requires the Secretaries, agencies and other interested parties to establish procedures to expedite applications for the construction of pipeline and transmission facilities over the designated corridors and

---

231 2005 Energy Policy Act § 366.

232 2005 Energy Policy Act § 367.

233 2005 Energy Policy Act § 367.

234 The Federal Land Policy and Management Act of 1976, 43 U.S.C. § 1702(o), defines the western states as the States of Arizona, California, Colorado, Idaho, Montana, Nevada, New Mexico, Oregon, Utah, Washington, and Wyoming.

235 2005 Energy Policy Act § 368(a).

236 2005 Energy Policy Act § 368(b).

to ensure additional corridors are promptly identified and designated as necessary.[237] The agencies must consider the need for upgraded and new electricity transmission and distribution facilities to improve reliability, relieve congestion and enhance the capability of the national grid to deliver electricity.[238]

To further these ambitious objectives, the Act requires the Secretaries of Energy, Interior, Agriculture and Defense to enter into an MOU to facilitate the designation of corridors for oil, gas, and hydrogen pipelines and electricity transmission and distribution facilities on federal land.[239] The MOU is intended to coordinate all applicable federal authorizations and environmental reviews related to a proposed or existing utility facility.[240] The MOU is to include provisions that (1) establish a unified right-of-way application form and administrative procedure for processing the applications, including lines of authority, application process steps, and a timeframe; (2) provide for coordinated planning in granting rights-of-way and use of right-of-way stipulations to achieve consistency; and (3) provide for an agreement among the affected federal agencies to prepare a single environmental review document to be used as the basis for all federal authorization decisions.[241]

---

[237] 2005 Energy Policy Act § 368(c).

[238] 2005 Energy Policy Act § 368(d).

[239] 2005 Energy Policy Act § 372.

[240] 2005 Energy Policy Act § 372(a)(1). The Act defines a "utility facility" as

> any privately, publicly, or cooperatively owned line, facility, or system—
>
> (1) for the transportation of—
>
> (A) oil, natural gas, synthetic liquid fuel, or gaseous fuel;
>
> (B) any refined product produced from oil, natural gas, synthetic liquid fuel, or gaseous fuel; or
>
> (C) products in support of the production of material referred to in subparagraph (A) or (B);
>
> (2) for storage and terminal facilities in connection with the production of material referred to in paragraph (1); or
>
> (3) for the generation, transmission, and distribution of electric energy.

2005 Energy Policy Act § 372(c).

[241] 2005 Energy Policy Act § 372(c).

[i] Oil Shale, Tar Sands, and Other Strategic Unconventional Fuels Act of 2005

Section 369 of the Act enacts the Oil Shale, Tar Sands, and Other Strategic Unconventional Fuels Act of 2005 (Unconventional Fuels Act or UFA). Congress declared that oil shale, tar sands and other unconventional fuels are strategically important resources that should be developed to reduce dependence on foreign oil supplies. Therefore, the UFA establishes a leasing program for research and development of oil shale and tar sands,[242] and requires the completion of a programmatic environmental impact statement for a commercial leasing program for oil shale and tar sands on public lands with an emphasis on the most geologically prospective lands within Colorado, Utah and Wyoming.[243] Within six months after the Secretary of the Interior completes the study, the Secretary must issue new regulations establishing a leasing program. Within 180 days of publication of the new regulations, the Secretary must, in consultation with others, evaluate the interest of a state in developing oil shale and tar sand resources.[244] The Act authorizes the Secretary of the Interior to conduct a lease sale of public land with oil shale and tar sands subject to sufficient support and interest in a state.[245]

The UFA delegates a variety of functions and responsibilities to federal agencies. The Office of Petroleum Reserves under the Department of Energy is charged with the primary tasks of creating and implementing a commercial strategic fuel development program, evaluating the strategic importance of unconventional sources of strategic fuels to the security of the United States, and promoting and coordinating federal government actions to develop strategic fuels to address the energy supply needs of the United States. The Office of Petroleum Reserves will also work with a Task Force composed of the Secretaries of Energy, Interior, and Defense, the governors of affected states, and representatives of local governments in affected areas to provide, in part, recommendations on strategic unconventional fuel development.[246] The Department of the Interior will serve as the

---

242 2005 Energy Policy Act § 369(c).

243 2005 Energy Policy Act § 369(d).

244 2005 Energy Policy Act § 369(d)(1).

245 2005 Energy Policy Act § 369(e).

246 2005 Energy Policy Act § 369(h).

lead agency for coordination of the permit review process,[247] carry out an assessment of oil shale and tar sand resources in designated geographic areas,[248] facilitate land exchanges to recover oil shale and tar sands,[249] identify and prioritize public lands containing deposits of oil shale and tar sands, and establish royalty rates for leases.[250] The Department of Energy is charged with identifying technologies for the development of oil shale and tar sands, including possible technical, regulatory compliance and cost-sharing assistance for those technologies identified[251] and updating the 1987 technical and economic assessment of domestic heavy oil resources prepared by the Interstate Oil and Gas Compact Commission.[252] The Department of Defense is authorized to enter into agreements to procure fuel produced from coal, oil shale and tar sands (defined as "covered fuels"), if clean fuel requirements can be met, and to develop a strategy for using covered fuels.[253] The UFA also authorizes appropriations to carry out its provisions, as necessary.[254]

[j] Finger Lakes Withdrawal

The Act withdraws all federal land within the Finger Lakes National Forest in the State of New York from all forms of entry, appropriation, or disposal under public land laws and from disposition under all laws relating to oil and gas leasing.[255]

---

247 2005 Energy Policy Act § 369(k).

248 2005 Energy Policy Act § 369(m). The geographic areas are listed in order of priority in § 369(m)(1)(B) as follows: the Green River Region of Colorado, Utah and Wyoming; the Devonian oil shales and other hydrocarbon-bearing rocks having the nomenclature of "shale" located east of the Mississippi River; and any remaining area in the central and western U.S., including Alaska, that contains oil shale and tar sands.

249 2005 Energy Policy Act § 369(n).

250 2005 Energy Policy Act § 369(o).

251 2005 Energy Policy Act § 369(l).

252 2005 Energy Policy Act § 369(p).

253 2005 Energy Policy Act § 369(q).

254 2005 Energy Policy Act § 369(s).

255 2005 Energy Policy Act § 370.

[k] Reinstatement of Leases

The Act implements two adjustments to the provisions under the Mineral Leasing Act[256] pertaining to reinstatement of leases terminated for failure to pay the full amount of rental on or before the anniversary date of the lease.[257] First, it provides a 120-day window following enactment of the Act for oil and gas lessees to file a petition to have their lease reinstated where the lease was terminated for failure of a lessee to pay the full amount of rental on or before the anniversary date of the lease during the period beginning on September 1, 2001, and ending on June 30, 2004.[258] The lessee must file a petition for reinstatement and comply with the necessary requirements, most notably payment of back leases and back royalties.[259] Second, the Act amends § 31(d)(2) of the Mineral Leasing Act to extend the period of time for lessees to petition for reinstatement of leases terminated *after* enactment of the Act. For those petitions, the petition must be filed on or before the earlier of (1) 60 days after receipt of the termination notice, or (2) 24 months after termination of the lease.[260] For petitions on leases terminated *before* enactment of the Act and that are not filed in the 120-day window specified above, the petition must be filed on or before the earlier of (1) 60 days after receipt of the termination notice, or (2) 15 months after termination of the lease.[261]

[l] Sense of Congress Regarding Development of Minerals Under Padre Island National Seashore

The Act clarifies Congress' understanding that the United States owns only the surface estate of certain lands constituting the Padre Island National Seashore and that ownership of the oil, gas, and other mineral interests in the subsurface estate are held by the State of Texas and private parties.[262]

---

[256] 30 U.S.C. § 188(d)(2).

[257] 2005 Energy Policy Act § 371.

[258] 2005 Energy Policy Act § 371(a).

[259] Conditions for reinstatement are set forth in 30 U.S.C. § 188(e).

[260] 2005 Energy Policy Act § 371(b).

[261] The 60 day/15 month petition filing period for lease terminations prior to enactment of the Act follows the filing period in place under Section 31(d)(2) of the Mineral Leasing Act, 30 U.S.C. 188(d)(2) prior to its amendment under the Act.

[262] 2005 Energy Policy Act § 373.

[m] Livingston Parish Mineral Rights Transfer

Section 102 of Public Law 102-562 was enacted to resolve adverse claims and title confusion arising from certain private land claims in Livingston Parish, Louisiana predating the Louisiana Purchase.[263] At that time, title was conveyed with exception and reservation of all mineral rights in the land. Section 374 of the Act requires that, within one year of enactment, the United States Geological Survey shall conduct and publish an assessment of all oil and gas resources underlying certain lands in Livingston Parish.[264] Upon a finding by the Secretary of the Interior that it is unlikely that economically recoverable oil and gas resources are present on the lands, the Secretary shall convey all rights to oil and gas within 180 days.[265] All remaining mineral rights are to be conveyed under the Act to all holders of a right, title, or interest to such mineral rights under Louisiana law no later than 180 days after enactment of the Act.[266]

[8] Miscellaneous Provisions Under Title III—Oil and Gas

[a] Deadline for Decision on Appeals of Consistency Determination Under the Coastal Zone Management Act of 1972

Section 381 of the Act amends the Coastal Zone Management Act of 1972 (CZMA)[267] to shorten the time period for issuing a decision in the appeal of a consistency determination to the Secretary of Commerce.[268] Prior to the amendment, the Secretary had up to 90 days from publication of the notice closing the decisional record to issue a final decision in the appeal or publish a notice in the Federal Register detailing why a decision could not be issued within the 90-day period. Under the amended CZMA, the time period is shortened to 60 days.[269] Additionally, if a notice was published explaining why a decision was not issued, the CZMA previously provided the Secretary up to 45 days from the date of publication to issue the decision.

---

263 Pub. L. No. 102-562 § 102, 106 Stat. 4234 (1992).

264 2005 Energy Policy Act § 374.

265 2005 Energy Policy Act § 374.

266 2005 Energy Policy Act § 374.

267 16 U.S.C. § 1465.

268 2005 Energy Policy Act § 381.

269 2005 Energy Policy Act § 381(c)(1).

Under the amended CZMA, that time period is shortened to 15 days.[270] The new provisions also provide for a stay to the closing of the decisional record.[271]

[b] Appeals Relating to Offshore Mineral Development

The Act provides that, for any federal administrative agency proceeding that is an appeal or review under Section 319 of the CZMA[272] related to federal authorization of an energy project, the lead federal permitting agency for the energy project will maintain a consolidated record of decisions made and actions taken by the lead agency or other federal and state administrative agencies. The consolidated record will serve as the initial record for the Section 319 appeal.[273]

[c] Royalty Payments Under Leases Under the Outer Continental Shelf Lands Act

The Oil Pollution Act of 1990[274] provided compensation in Section 6004(c) to the State of Louisiana and lessees for net drainage of oil and gas resources in certain leases. Section 383 of the Act provides royalty relief for Section 6004(c) lessees with leases issued under the Outer Continental Shelf Lands Act.[275] A lessee with royalty payments due and payable in the period beginning on October 1, 2006, may withhold from payment any royalty due and owing to the United States under the Outer Continental Shelf Lands Act for offshore oil or gas production from a covered lease tract[276] if, on or before the date that the payment is due, the lessee makes a payment of 44 cents for every $1 of royalty withheld.[277] The lessee's relief from payment of royalties

---

[270] 2005 Energy Policy Act § 381(c)(2).

[271] 2005 Energy Policy Act § 381(b)(3).

[272] 16 U.S.C. § 1465.

[273] 2005 Energy Policy Act § 382.

[274] Pub. L. No. 101-380, 104 Stat. 484 (1990).

[275] 43 U.S.C. § 1301 *et seq.*

[276] "The term 'covered lease tract' means a leased tract (or portion of a leased tract)— (A) lying seaward of the zone defined and governed by section 8(g) of the Outer Continental Shelf Lands Act (43 U.S.C. § 1337(g)); or (B) lying within such zone but to which such section does not apply." 2005 Energy Policy Act § 383(c)(1).

[277] 2005 Energy Policy Act § 383(a)–(b).

ends on the date that the Secretary of the Treasury publishes a notice that the total amount of the royalty withheld by the lessee equals certain pre-determined dollar amounts, plus interest.[278]

[d] Coastal Impact Assistance Program

The Act amends the Coastal Impact Assistance Program under Section 31 of the Outer Continental Shelf Lands Act.[279] The Program provides funding to Producing States[280] and Coastal Political Subdivisions[281] from qualified Outer Continental Shelf revenues[282] collected from oil and natural gas activities on the Outer Continental Shelf. The Act amends the payment provision to provide appropriations of $250 million to Producing States and Coastal Political Subdivisions for each of fiscal years 2007-2010.[283] The funding provided under the Program has limited use. Under the Act, Program funds may be used for coastal and wetland conservation, protection or restoration projects; mitigation

---

278 2005 Energy Policy Act § 383(b).

279 43 U.S.C. § 1356a.

280 The Act defines a "Producing State" as a "coastal State that has a seaward boundary within 200 nautical miles of the geographic center of a leased tract within any area of the outer Continental Shelf," except for Producing States in which "a majority of the coastline is subject to a leasing moratoria, unless production was occurring on January 1, 2005, from a lease within 10 nautical miles of the coastline of that State." 2005 Energy Policy Act § 384.

281 The Act defines a "Coastal Political Subdivison" as "a political subdivision of a coastal State any part of which political subdivision is within the coastal zone (as defined in Section 304 of the Coastal Zone Management Act of 1972 (16 U.S.C. § 1453)) of the coastal State as of the date of enactment of the Energy Policy Act of 2005; and not more than 200 nautical miles from the geographic center of any leased tract." 2005 Energy Policy Act § 384.

282 The Act defines "qualified Outer Continental Shelf revenues" to mean "all amounts received by the United States from each leased tract or portion of a leased tract" within a zone designated under the Act from "bonus bids, rents, royalties (including payments for royalty taken in kind and sold), net profit share payments, and related late-payment interest from natural gas and oil leases issued under this Act." 2005 Energy Policy Act § 384.

283 2005 Energy Policy Act § 384. The Act allocates the funds among the eligible states under this program by focusing first on a ratio of the amount of revenue generated from the Outer Continental Shelf production associated with a state compared to the overall revenue generated from Outer Continental Shelf production. *See* 2005 Energy Policy Act § 384. This amount is further allocated to each state's Political Subdivisions in varying amounts. *See* 2005 Energy Policy Act § 384.

of damages to fish, wildlife or natural resources; marine, coastal or conservation management plans; mitigation of impacts on the Outer Continental Shelf on onshore infrastructure and public service needs; and associated planning assistance and administrative costs.[284] To receive disbursements, each state governor must submit to the Secretary of the Interior a coastal state assistance plan. The plan is to include input from public participation and outline how Program funding will be used, the state agency acting as representative on behalf of the Program, implementation measures for use of the funds, and a contact person and description of funding uses in Coastal Political Subdivisions.[285]

[e] Study of Availability of Skilled Workers

The Act requires the Secretary of Energy to enter into an arrangement with the National Academy of Sciences to conduct a study of the short-term and long-term availability of skilled workers to meet the energy and mineral security requirements of the United States.[286] The study is to include an analysis of the need for and availability of workers for the oil, gas and mineral industries, the availability of skilled workers at entry and senior levels, and recommendations for action to meet future labor requirements.[287] The Secretary of Energy is to issue a report on the results of the study within two years of the enactment of the Act.[288]

[f] Great Lakes Oil and Gas Drilling Ban

The Act prohibits any federal or state permit or lease to be issued for new oil and gas slant, directional, or offshore drilling in or under one or more of the Great Lakes.[289]

[g] Federal Coalbed Methane Regulation

The Energy Policy Act of 1992 set forth a list of states, referred to as the Affected States, in which either disputes exist over the

284 2005 Energy Policy Act § 384.

285 2005 Energy Policy Act § 384.

286 2005 Energy Policy Act § 385.

287 2005 Energy Policy Act § 385.

288 2005 Energy Policy Act § 385.

289 2005 Energy Policy Act § 386.

ownership rights of coalbed methane or various hindrances slow the development of coalbed methane gas. The Act provides that any state currently on the list of Affected States established under Section 13368(b) of the Energy Policy Act of 1992[290] shall be removed from the list if, not later than three years after the date of enactment of this Act, the state takes or has taken prior to the date of enactment, any of the actions required for removal as set forth under Section 1339(b).[291]

[h] Alternate Energy-Related Uses on the Outer Continental Shelf

The Act adds a provision to the Outer Continental Shelf Lands Act[292] providing the Secretary of Energy with authority to grant a lease, easement or right-of-way on the Outer Continental Shelf for activities not otherwise authorized under this Act, the Deepwater Port Act of 1974,[293] the Ocean Thermal Energy Conversion Act of 1980,[294] or other applicable law. With the exception of areas where activity is prohibited by a moratorium, lease activities must either (1) support exploration, development, production or storage of oil or natural gas; (2) support transportation of oil or natural gas, excluding shipping activities; (3) produce or support production, transportation or transmission of energy from sources other than oil and gas; or (4) use, for energy-related purposes or for other authorized marine-related purposes, facilities currently or previously used for activities authorized under the Act.[295] The Secretary of Energy establishes the royalties and payments for any lease, easement or right-of-way,[296] as well as providing coordination and consultation with affected state and local communities.

The Act also provides for a coordinated Outer Continental Shelf mapping initiative. The initiative is to be coordinated with several

290 42 U.S.C. § 13368(b).

291 2005 Energy Policy Act § 387.

292 43 U.S.C. § 1337.

293 33 U.S.C. § 1501 *et seq.*

294 42 U.S.C. § 9101 *et seq.*

295 2005 Energy Policy Act § 388(a).

296 2005 Energy Policy Act § 388(a)(2).

federal agencies and designed to establish an interagency, comprehensive digital mapping initiative. The maps should include indications on the Outer Continental Shelf of federally-permitted activities; obstructions to navigation; submerged cultural resources; undersea cables; offshore aquaculture products; and any area designated for the purpose of safety and national security, environmental protection, or conservation and management of living marine resources.[297]

[i] Oil Spill Recovery Institute

The Act extends funding for the Oil Spill Recovery Institute. Previously, the funding termination provisions in the Oil Pollution Act of 1990[298] triggered termination of funding with a set date. The provisions were re-written so that funding terminates "1 year after the date on which the Secretary, in consultation with the Secretary of the Interior, determines that oil and gas exploration, development, and production in the State of Alaska have ceased."[299]

[j] NEPA Review

The Act provides five categorical exclusions to the NEPA for activities conducted pursuant to the Mineral Leasing Act for the purpose of exploration and development of oil or gas. The excluded activities include (1) individual surface disturbances of less than five acres so long as the total surface disturbance on the lease is not greater than 150 acres and site-specific analysis in documents prepared pursuant to the NEPA has been previously completed; (2) drilling an oil or gas well at a location or well pad site where drilling has occurred previously within the past five years prior to the date of spudding in the well; (3) drilling an oil or gas well within a developed field for which an approved land use plan or any environmental document prepared pursuant to the NEPA analyzed such drilling and was approved within the past five years prior to the date of spudding in the well; (4) placement of a pipeline in an approved right-of-way corridor where the corridor was approved within the five years prior to the date of placement; and (5) maintenance of a minor activity, other than any construction or major renovation of a building or facility.[300]

---

297 2005 Energy Policy Act § 388(b).

298 33 U.S.C. § 2731 *et seq.*

299 2005 Energy Policy Act § 389.

300 2005 Energy Policy Act § 390.

[9] Refinery Revitalization

The Act sets forth several findings on the current supply and exceeding demand for refinery capacity in the United States. It also provides authority for the EPA Administrator to enter into refining permitting cooperative agreements with the governor of a state to streamline the environmental permitting process and assist states in hiring personnel to complete state permitting reviews.[301]

[a] Findings and Definitions

Under the Act, Congress concludes that it is in the national interest to increase petroleum refining capacity for gasoline, heating oil, and various petrochemical products to bring more supply to the market.[302] Demand for refined petroleum products currently exceeds the supply, resulting in increased imports projected to grow from 7.9% to 10.7% by 2025.[303] Refiners are subject to significant environmental and other regulatory requirements and face several new requirements over the next decade, including requirements under the Clean Air Act.[304] The Act calls for better coordination between federal and state regulatory agencies for siting, permitting and construction of new refineries in an effort to meet the demand in the United States for refined products.[305]

[b] Federal-State Regulatory Coordination and Assistance

At the request of a governor of a state, the Act authorizes the EPA Administrator to enter into a cooperative agreement with the state for permitting of a new refinery.[306] The agreement is to recognize steps, including a timeline, that are needed to streamline consideration of federal and state environmental permits for the new facility. The Administrator is authorized to accept a consolidated application from the refiner for all permits required from the EPA, enter in a Memorandum of Understanding with other federal agencies to coordinate consideration of refinery applications and permits for the facility, and

---

301 2005 Energy Policy Act § 392.

302 2005 Energy Policy Act § 391.

303 2005 Energy Policy Act § 390.

304 42 U.S.C. § 7401 *et seq.*

305 2005 Energy Policy Act § 391.

306 2005 Energy Policy Act § 392.

enter into a Memorandum of Understanding with the state to coordinate concurrent review of federal and state permit applications.[307] The Administrator also has authority to provide technical and legal assistance to the state, as well as financial assistance to hire additional personnel in relevant fields to assist in consideration of refinery permits.[308]

§ 1.04 Title IV—Coal

[1] Scope

Coal is by far the nation's most abundant native fossil fuel source and occupies a central role in the energy industry. Over 90% of the coal used in the United States is used to generate electricity.[1] About 50% of the electricity generated in the United States comes from coal-fired plants, compared to the approximately 19% generated by nuclear plants and 18% by gas-fired plants.[2] Given its prevalence and its chemical and physical properties, coal presents unique environmental issues as well. In recognition of these considerations, Title IV of the Act contains provisions seeking both to foster the usage of new, cleaner technologies for coal usage and to increase production through expanded leasing rights on federal lands and other incentives.

[2] Establishment of Clean Coal Power Initiative (CCPI)

The Act establishes a Clean Coal Power Initiative (CCPI), providing funding and loan guarantees for clean coal technology that promotes efficient, environment-friendly, and cost-competitive coal technologies.[3] The Secretary of Energy administers the CCPI program, for

---

[307] 2005 Energy Policy Act § 392(b).

[308] 2005 Energy Policy Act § 392(c)–(d).

[1] Energy Information Administration, Annual Coal Report 2004 at 5 (published November 2005) ("The electric power sector (electric utilities and independent power producers) accounted for almost 92 percent of all coal consumed in the United States in 2004.").

[2] Energy Information Administration, Electric Power Monthly, October 2005, at 1 ("Year-to-date through July 2005, 49.6 percent of the Nation's electric power was generated at coal-fired plants . . . . Nuclear plants contributed 19.3 percent of the total generation, while natural gas-fired plants generated 18.4 percent, and 2.8 percent was generated at petroleum-fired plants. Conventional hydroelectric power provided 7.3 percent of the total, while other renewables and other miscellaneous energy sources generated the remaining electric power.).

[3] 2005 Energy Policy Act § 401.

which the Act authorizes $200 million annually, for fiscal years 2006-2015.[4] The CCPI program is intended to build upon the Department of Energy's (DOE) preexisting Clean Coal Technology program, which began in 1986 and already had a number of successes.[5] That predecessor program began as a response to concerns over acid rain, caused by the sulfur and nitrogen pollutants sometimes emitted by coal-burning power plants. Special Envoys of the United States and Canadian governments recommended to President Ronald Reagan that the Clean Coal Technology Program be established and jointly funded by the United States government, state agencies, and the private sector.[6] The program is recognized for many successes, such as the low nitrogen oxide (NOx) burner technology developed under the Clean Coal Technology Program, which is now found on 75% of United States coal fired power plants.[7] Also developed under the program was the modern scrubber technology that has reduced capital and operating costs for sulfur removal to about half of the previous level.[8]

[3] General Criteria for CCPI Projects

The Act establishes a high standard for projects to qualify for CCPI funding. To be eligible, a project must "advance efficiency, environmental performance, and cost competitiveness."[9] Moreover, the project must be one that achieves these benefits "well beyond the level of technologies that are in commercial service or have been demonstrated on a scale that the Secretary determines is sufficient to demonstrate that commercial service is viable as of the date of

[4] 2005 Energy Policy Act § 401(a).

[5] DOE website, "The 1986-93 Clean Coal Technology Program."

[6] DOE website, "The 1986-93 Clean Coal Technology Program."

[7] Statement of John E. Shelk, Senior Vice President, Government Affairs, National Mining Association, before the House Committee on Energy and Commerce Subcommittee on Energy and Air Quality, February 16, 2005, available at http://energycommerce.house.gov/108/Hearings/02162005hearing1437/Shelk.pdf.

[8] Statement of John E. Shelk, Senior Vice President, Government Affairs, National Mining Association, before the House Committee on Energy and Commerce Subcommittee on Energy and Air Quality, February 16, 2005, available at http://energycommerce.house.gov/108/Hearings/02162005hearing1437/Shelk.pdf.

[9] 2005 Energy Policy Act § 402(a).

enactment of this Act."[10] Technical criteria, discussed below, apply as well.

[4] CCPI Funding of Coal Gasification Projects

The primary emphasis of the CCPI program is on coal gasification, a process that converts coal from a solid to a gaseous state. Gas produced through coal gasification is very similar to natural gas and can be used to make chemicals or fertilizer or, more relevant to CCPI, electric power. At least 70% of the amounts expended by DOE in the program must be used for the funding of coal-based gasification technologies.[11] The Act specifies that such technologies are to include gasification combined cycle, gasification fuel cells and turbine combined cycle, gasification coproduction, hybrid gasification and combustion, and other advanced coal based technologies capable of producing a concentrated stream of carbon dioxide.[12] The DOE, under the Act, is to establish technical milestones specifying the emission and thermal efficiency levels that coal gasification projects under this subtitle shall be designed, and reasonably expected, to achieve.[13] During the 10-year period of the CCPI program, the DOE's technical milestones for coal gasification are to be "more prescriptive" and are intended to ensure that the coal gasification industry reaches a specified finish line by 2020: that year is the statutory deadline for coal gasification projects in the CCPI program to be capable of removing at least 99% of sulfur dioxide; emitting not more than 0.05 pounds of NOx per million Btu; achieving at least 95% reductions in mercury emissions; and achieving a thermal efficiency of at least (1) 50% for coal with a heat value of more than 9,000 Btu (per pound), (2) 48% for coal of 7,000 to 9,000 Btu, and (3) 46% for coal of less than 7,000 Btu.[14]

[5] CCPI Funding of Other (Non-Gasification) Projects

As noted above, 70% or more of the CCPI's $200 million per year is to be spent on coal gasification. The balance may be allocated to

---

[10] 2005 Energy Policy Act § 402(a).

[11] 2005 Energy Policy Act § 402(b)(1)(A).

[12] 2005 Energy Policy Act § 402(b)(1)(A).

[13] 2005 Energy Policy Act § 402(b)(1)(B).

[14] 2005 Energy Policy Act § 402(b)(1)(B).

other clean coal projects. As the DOE is required to do for gasification projects, for non-gasification projects the DOE is to issue periodic technical milestones specifying the emission and thermal efficiency levels that the projects are to be designed, and reasonably expected, to achieve.[15] In the case of this latter category of projects, however, the performance standards to be set by the DOE for 2020 are slightly less ambitious; the Act requires DOE to set the periodic milestones so as to achieve by the year 2020 projects capable of (1) removing at least 97% of sulfur dioxide; (2) emitting not more than 0.08 pounds of NOx per million Btu; (3) achieving at least 90% reductions in mercury emissions; and (4) achieving a thermal efficiency of at least (a) 43% for coal of more than 9,000 Btu (per pound), (b) 41% for coal of 7,000 to 9,000 Btu, and (c) 39% for coal of less than 7,000 Btu.[16]

[6] CCPI Prescriptive Milestone Consultation and Evaluation

The DOE, prior to establishing the periodic technical milestones contemplated by the Act's CCPI provisions, is obligated to engage in consultation to ensure that the milestones are realistic and reflect input from a wide range of perspectives. The Act requires that the DOE's consultations include the Administrator of the Environmental Protection Agency and interested entities including coal producers, industries using coal, organizations that promote coal or advanced coal technologies, environmental organizations, organizations representing workers, and organizations representing consumers.[17] For CCPI projects at facilities that already were in existence at the time of enactment of the Act, the standards are more lenient; the milestones applicable to such projects are to be designed to achieve an overall thermal design efficiency *improvement*, compared to the efficiency of the unit as operated previously, of not less than (1) 7% for coal of more than 9,000 Btu (per pound), (2) 6% for coal of 7,000 to 9,000 Btu, and (3) 4% for coal of less than 7,000 Btu. In evaluating a proposed project's ability to satisfy the technical milestone standards, the DOE is obligated to take into account the project site's altitude

[15] 2005 Energy Policy Act § 402(b)(2)(B).

[16] 2005 Energy Policy Act § 402(b)(2)(B)(ii).

[17] 2005 Energy Policy Act § 402(b)(3).

and to make adjustments to the milestones as applied to the project, where warranted by altitude. The DOE is permitted to give priority to proposed projects involving the separation or capture of carbon dioxide or the reduction of the demand for natural gas if the project is deployed.[18]

[7] Financial Criteria for CCPI Projects

Under the CCPI program, the DOE is prohibited from providing any financial assistance to a proposed project unless the recipient demonstrates to the satisfaction of the Secretary that (1) the recipient is financially responsible; (2) the recipient will provide sufficient information to the Secretary to enable the Secretary to ensure that the funds are spent efficiently and effectively; and (3) a market exists for the technology being demonstrated or applied, as evidenced by statements of interest in writing from potential purchasers of the technology.[19]

[8] DOE Awards of Assistance Under CCPI

If a proposed CCPI project is able to demonstrate that it satisfies the objective criteria outlined above, then, pursuant to the Act, the DOE is obligated to provide financial assistance to the project, provided that the DOE finds that the project is likely (1) to achieve overall cost reductions in the use of coal to generate useful forms of energy or chemical feedstocks; (2) to improve the competitiveness of coal among various forms of energy in order to maintain a diversity of fuel choices in the United States to meet electricity generation requirements; and (3) to demonstrate methods and equipment that are applicable to 25% of the electricity generating facilities, using various types of coal, that use coal as the primary feedstock as of the date of enactment of this Act.[20] Where the DOE decides to grant funding to a particular CCPI project, it must establish a reasonable period of time during which the owner or operator of the project is obligated to complete the construction or demonstration phase of the project.[21] Awards of CCPI funding are expressly conditioned on the recipient's

[18] 2005 Energy Policy Act § 402(b)(5).

[19] 2005 Energy Policy Act § 402(c).

[20] 2005 Energy Policy Act § 402(d).

[21] 2005 Energy Policy Act § 402(f).

commitment not to seek any extension of the time period established by the DOE.[22] The DOE is permitted, in its sole discretion, to extend a CCPI project's completion deadline, but may not grant an extension greater than four years.[23] The DOE is further permitted to vest fee title or other property interests acquired under CCPI agreements in any entity, including the United States.[24]

[9] CCPI Awards Subject to Cost-Sharing Requirements

All awards of financial assistance by the DOE to CCPI projects under the Act are subject to the Act's provisions on cost-sharing, set forth in Section 988 of the statute.[25] Thus, for CCPI projects, which are aimed at commercial applications, the non-federal funding sources for a project must be at least 50%.[26] The DOE has the authority to reduce that 50% threshold upon a finding that the reduction is necessary and appropriate, taking into consideration any technological risk relating to the activity.[27]

[10] Data Protection for CCPI Projects

For up to five years after the operations phase of a project under a CCPI cooperative agreement, the DOE is permitted to provide appropriate protections (including FOIA exemptions) against the dissemination of information that (1) results from demonstration activities carried out under the CCPI, and (2) would be a trade secret or commercial or financial information that is privileged or confidential if the information had been obtained from and first produced by a non-federal party participating in a CCPI project.[28]

[11] Report to Congress on CCPI

Within a year after the date of enactment of the Act, and once every two years thereafter through 2014, the DOE is obligated to report to Congress on the experience under the CCPI program. The report must

---

22 2005 Energy Policy Act § 402(f)(2).

23 2005 Energy Policy Act § 402(f)(3).

24 2005 Energy Policy Act § 402(g).

25 2005 Energy Policy Act § 988.

26 2005 Energy Policy Act § 988(c).

27 2005 Energy Policy Act § 988(c)(2).

28 2005 Energy Policy Act § 402(h).

describe (1) the periodic technical milestones that have been issued to date, and how those milestones ensure progress toward meeting the relevant statutory goals, and (2) the status of all CCPI projects funded to date.[29]

[12] Clean Coal Centers of Excellence

The DOE, as part of the CCPI program, is to award competitive, merit-based grants to institutions of higher education for the establishment of "Clean Coal Centers of Excellence" — academic programs for studying and planning for the energy systems of the future. In making its grant selections, the DOE is to select those institutions of higher education that show the greatest potential for advancing new clean coal technologies.[30]

[13] Integrated Coal/Renewable Energy Systems

The DOE is authorized to provide loan guarantees, subject to appropriations limitations, for projects to produce energy from coal having a heat value of less than 7,000 Btu (per pound) using advanced integrated gasification combined cycle (IGCC) technology. For a project to receive a loan guarantee under this program, the coal-burning elements of the project must be combined with wind or other renewables. The project also must minimize carbon dioxide emissions and provide the potential to sequester such emissions, and it must provide a ready source of hydrogen for near-site fuel cell demonstrations.[31] This program is limited to projects located in the Upper Great Plains.[32]

[14] Healy, Alaska Clean Coal Technology Plant

The Act contains a special provision providing for the issuance of a loan of up to $80 million for an experimental plant located in Healy, Alaska, constructed under a DOE cooperative agreement. The terms and conditions for the loan are specified in the language of the Act.[33]

---

[29] 2005 Energy Policy Act § 403.

[30] 2005 Energy Policy Act § 404.

[31] 2005 Energy Policy Act § 411.

[32] 2005 Energy Policy Act § 411.

[33] 2005 Energy Policy Act § 412.

[15] Western Integrated Coal Gasification Demonstration Project

Subject to the availability of appropriations, the DOE is to establish a project to demonstrate production of energy from coal mined in the western United States using IGCC technology. The demonstration project is permitted to include repowering of existing facilities. The project is to be designed to demonstrate the ability to make use of coal with an energy content of not more than 9,000 Btu per pound, along with other types of coal (including subbituminous and bituminous coal with an energy content of up to 13,000 Btu), mined in the western United States. The Act requires that the demonstration project be located in a western state at an altitude of greater than 4,000 feet above sea level.[34]

[16] Coal Gasification

The Act authorizes the DOE to provide loan guarantees for a project to produce energy using IGCC technology. The project must be at least 400 MW and must produce power at competitive rates in deregulated energy generation. It also is barred from receiving any subsidy from ratepayers.[35]

[17] Petroleum Coke Gasification and Electron Scrubbing

Under the Act, the DOE is authorized to provide loan guarantees for at least five projects for the gasification of petroleum coke—a heavy residue that is a byproduct of petroleum processing.[36] The DOE also is to expend $5 million through its Chicago Operations office on a project to demonstrate the viability of high-energy electron scrubbing technology on commercial scale electrical generation using high-sulfur coal.[37]

[18] DOE Transportation Fuels From Illinois Basin Coal

The Act authorizes the DOE to establish a program to evaluate commercial and technical viability of advanced technologies for

---

[34] 2005 Energy Policy Act § 413.

[35] 2005 Energy Policy Act § 414.

[36] 2005 Energy Policy Act § 415.

[37] 2005 Energy Policy Act § 416.

production of Fischer-Tropsch transportation fuels and other transportation fuels from Illinois basin coal. Specific test facilities are authorized to be constructed at Southern Illinois, Kentucky, and Purdue Universities. The Act likewise authorizes construction of a test center for coal gasification products. For these activities, the Act authorizes expenditures of $85 million for the period of fiscal years 2006-2010.

[19] Clean Air Coal Program

The Act promotes the increased use of coal, particularly for purposes of electricity generation. Section 421 of the Act amends the Energy Policy Act of 1992 to add a new title, "Clean Air Coal Program" (CACP).[38] The stated purposes of the new CACP title are to "(1) promote national energy policy and energy security, diversity, and economic competitiveness benefits that result from the increased use of coal; (2) mitigate financial risks, reduce the cost of clean coal generation, and increase the marketplace acceptance of clean coal generation and pollution control equipment and processes; and (3) facilitate the environmental performance of clean coal generation."[39] A separate provision of the new title spells out more plainly the strongly pro-coal nature of the CACP; the program provides financial assistance to "facilitate the production and generation of coal-based power, through the deployment of clean coal electric generating equipment and processes" that improve energy efficiency or environmental performance consistent with relevant federal and state clean air requirements.[40] Another basis for obtaining CACP funds is to "facilitate the utilization of existing coal-based electricity generation plants through projects that (A) deploy advanced air pollution control equipment and processes; and (B) are designed to voluntarily enhance environmental performance above current applicable obligations under the Clean Air Act and State implementation efforts pursuant to such Act."[41]

[20] Generation Projects Under CACP

Under the new CACP title of the Energy Policy Act of 1992, funding is available for generation projects. The Act specifies that the

[38] 2005 Energy Policy Act § 421.

[39] 2005 Energy Policy Act § 421.

[40] 2005 Energy Policy Act § 421.

[41] 2005 Energy Policy Act § 421.

funding is available for projects including (1) equipment or processes previously supported by a DOE program; (2) advanced combustion equipment and processes that the DOE determines will be cost-effective and could substantially contribute to meeting environmental or energy needs, including gasification, gasification fuel cells, gasification coproduction, oxidation combustion techniques, specialized facilities known as "ultra-supercritical boilers," and chemical looping; and (3) hybrid gasification/combustion systems, including systems integrating fuel cells with gasification or combustion units. The DOE is obligated under the Act to develop performance criteria for such projects, including specified thermal efficiency standards based on higher heating values for coal with different heat rates, for identified periods between 2006 and 2013. Between 25% and 75% of the projects must solely generate electricity, and a priority is assigned to technologies which are developed and demonstrated but not yet cost competitive, and which achieve greater efficiency and environmental performance. The Act authorizes $2.5 billion over seven years for such projects.[42]

[21] Air Quality Enhancement Program Under CACP

The CACP also establishes an Air Quality Enhancement Program, providing for cost-shared awards for existing coal-fired power plants that use advanced pollution control equipment and processes designed to meet or exceed current federal and state clean air requirements. The Act assigns a priority to (1) projects likely to result in significant air quality improvements in nonattainment air quality areas or substantially reduce the emission level of criteria pollutants and mercury air emissions, (2) projects for pollution control that result in the mitigation or collection of more than one pollutant, and (3) projects designed to allow the use of the waste byproducts or other byproducts of the equipment.[43]

[22] Federal Coal Leasing Amendments

The Act amends Section 3 of the Mineral Leasing Act[44] with the effect of repealing the preexisting 160-acre limit on federal coal leases.

---

[42] 2005 Energy Policy Act § 421.

[43] 2005 Energy Policy Act § 421.

[44] 30 U.S.C. § 203.

Leaseholders are permitted to produced coal from additional coal lands or coal deposits contiguous or cornering to those embraced in the lease. As a result, coal producers with federal leases are able to make use of additional federal coal lands without competitive bidding and without being obligated to pay production royalties. The limitation on these new leasing rights is that the total area added by a coal lease modification under this provision is capped at the lesser of 960 acres or twice the original lease acreage.[45] The Act also permits the DOE to extend a mining plan beyond the current 40-year limit set forth in Section 2(d)(2) of the Mineral Leasing Act,[46] provided that the DOE has found that the longer period "will ensure the maximum economic recovery of a coal deposit" or "is in the interest of the orderly, efficient, or economic development of a coal resource."[47]

[23] Revisions to Requirements Governing Advance Royalty Payments

A federal coal lease has an initial term of 20 years, and that term can be extended as long as coal is annually produced in commercial quantities, defined in Bureau of Land Management (BLM) regulations as 1% of the recoverable coal reserve under the lease or logical mining unit.[48] The requirement for annual production of commercial quantities of coal after satisfying diligent development requirements is commonly known as the continued operation requirement. The Act increases, from 10 years to 20 years, the aggregate number of years during the period of any lease for which advance royalties may be accepted in lieu of the continued operation requirement. The Act also establishes a new method for calculating advance royalty payments, requiring the price of such payments to be based on (1) the average price in the spot market for sales of comparable coal from the same region during the last month of each applicable continued operation year; or (2) in the absence of a spot market for comparable coal from the same region, by using a comparable method established by the Secretary of the Interior to capture the commercial value of coal. The Act also requires advance royalty payments to be based on usage of

---

[45] 2005 Energy Policy Act § 432.

[46] 30 U.S.C. § 202a(2).

[47] 2005 Energy Policy Act § 433.

[48] 43 C.F.R. § 3480.0-5(a)(6).

the term "commercial quantities" in a manner defined by regulation by the Secretary of the Interior.[49] A further liberalization of lessee obligations in the Act is the elimination of what had been a three-year deadline for a lessee to submit a coal lease operation and reclamation plan.[50] Yet another liberalization lies in the Act's treatment of the bonus bid—the dollar amount offered by a potential lessee as consideration for receiving a lease.[51] The bonus bid must meet or exceed fair market value to be accepted by the BLM.[52] A successful bonus bid (i.e., the highest bid that is equal to or greater than fair market value) need not provide for full payment by the lessee of the bid amount up front; the lessee may elect to pay one fifth up front and to pay the balance in installments.[53] The Act clarifies that the Mineral Leasing Act does not require a surety bond or other financial assurance to be furnished in connection with such deferred bonus bids for coal leases.[54] However, if the lessee fails to pay any installment of a deferred bonus bid, the lease terminates and any prior bonus bid payments are forfeited.[55]

[24] Inventory Requirement

The DOE is obligated under the Act, in consultation with the Secretaries of Agriculture and the Interior to compile a written inventory of developable coal resources on public lands and any restrictions on development of those resources. The DOE is required to publish a report to Congress containing each of these inventories and updates. The inventory and report are required to itemize "Compliant" coal resources (with sulfur content of 1.0 pounds to 1.2 pounds

---

49 2005 Energy Policy Act § 434.

50 2005 Energy Policy Act § 435.

51 *See* "The Federal Coal Leasing System," Bureau of Land Management, Department of the Interior (Oct. 1997), available at www.blm.gov/nhp/efoia/wy/1998im/wy1998-008atch.pdf.

52 *See* "The Federal Coal Leasing System," Bureau of Land Management, Department of the Interior (Oct. 1997), available at www.blm.gov/nhp/efoia/wy/1998im/wy1998-008atch.pdf.

53 *See* "The Federal Coal Leasing System," Bureau of Land Management, Department of the Interior (Oct. 1997), available at www.blm.gov/nhp/efoia/wy/1998im/wy1998-008atch.pdf.

54 2005 Energy Policy Act § 436.

55 2005 Energy Policy Act § 436.

of sulfur dioxide per MMBtu) and "Supercompliant" coal resources (with less than 1.0 pound of sulfur dioxide per MMBtu).[56]

§ 1.05 Title V—Indian Energy

[1] Scope

Title V of the Act, titled the Indian Tribal Energy Development and Self-Determination Act of 2005, fosters energy development on tribal lands. Title V also establishes the Office of Indian Energy Policy and Programs to promote Indian tribal energy resource development through grants and loans.[1]

[2] Office of Indian Energy Policy and Programs

Section 502 of the Act adds Section 217 to the Department of Energy Organization Act to create an Office of Indian Energy Policy and Programs within the Department of Energy (DOE). The new Office is charged with coordinating and implementing programs to promote the development of natural resources by Indian tribes, enhance and strengthen Indian tribal energy and economic infrastructure, reduce or stabilize energy costs, and bring electrical service to Indian lands.[2]

[3] Indian Energy

Section 503 amends Title 26 of the Energy Policy Act of 1992 to require the Department of the Interior to assist tribes and tribal energy resource development organizations in developing energy resources through grants and low-interest loans.[3] Grants and low-interest loans are to be used to develop management and technical skills, to develop energy resources on Indian land, and to establish tribal environmental programs to support energy programs.[4]

---

[56] 2005 Energy Policy Act § 437.

[1] 2005 Energy Policy Act Title V.

[2] 42 U.S.C. § 7131 *et seq.*

[3] 2005 Energy Policy Act § 2601(11) defines tribal energy resource development organizations as "two or more entities, at least one of which is an Indian tribe, that has the written consent of the governing bodies of all Indian tribes participating in the organization to apply for a grant, loan, or other assistance under section 2602."

[4] 25 U.S.C. § 3501 *et seq.*

The Director of the DOE's new Office of Indian Energy Policy and Programs must establish programs and provide grants to aid Indian tribes or tribal energy resource development organizations to promote efficiency and conservation programs and to support acquisition of energy supplies by tribes.[5] The Director also is required to support research projects involving carbon sequestration on Indian lands. An appropriation of $20 million is authorized for the fiscal years 2006-2016.

The Act requires the DOE to establish a loan guarantee program under which the Secretary of Energy can provide loan guarantees up to "90 percent of the unpaid principal and interest due on any loan made to an Indian tribe for energy development."[6] The aggregate amount of outstanding loan guarantees cannot exceed $2 billion at one time. Within one year after enactment of the Act, the Secretary of Energy must submit a report to Congress concerning financing requirements of tribes for energy development on Indian land.

The Act provides a preference to businesses with Indian-owned or controlled majorities when a federal agency or department purchases energy products, including electricity.[7] However, Congress made clear that the government cannot pay more than market price or receive less than prevailing market terms and conditions when making such purchases.

The Secretary of the Interior is authorized to provide grants to tribes for the development of tribal energy resource inventories on Indian land and for related feasibility studies.[8] Grants may also be used by tribes, with the exception of all but one Alaskan tribe, to develop and enforce tribal laws to regulate energy resource development, to develop the technical infrastructure to protect the environment, and for related employee training. Native Corporations under the Alaska Native Claims Settlement Act[9] may use such grants to develop and implement corporate policies related to energy resource development.

The Act establishes procedures under which Indian tribes may enter into leases, business agreements, and grant rights-of-way to

[5] 2005 Energy Policy Act § 2602(b).

[6] 2005 Energy Policy Act § 2602(c).

[7] 2005 Energy Policy Act § 2602(d).

[8] 2005 Energy Policy Act § 2603.

[9] 43 U.S.C. § 1602.

third-parties so that they may develop energy resources on tribal lands.[10] To be able to enter into such arrangements, an Indian tribe must demonstrate its financial and technical capacity and receive approval of the Secretary of the Interior of a tribal energy resource agreement. After obtaining the authority, subsequent business arrangements a Tribe may enter into with a third party that are within certain term limits and are that are for projects to serve a facility on tribal lands need not be reviewed by the Secretary.

Once the Secretary of the Interior promulgates regulations, tribes may apply to the Secretary of the Interior for tribal energy resource agreements. The Secretary has 270 days to act on an agreement, provided that no revisions are made by the tribe after the initial application is filed. In approving a tribal energy resource agreement, the Secretary will consider several criteria, including a tribe's ability to regulate the development of its energy resources, the economic return to the tribe under a lease or other business arrangement, whether technical and environmental issues have been addressed, and whether the resource agreement provides for consultation with impacted states regarding off-reservation effects. Tribal energy resource agreements must provide that the business arrangements will become effective on the date that a copy of the executed lease, business agreement or right-of-way is provided to the Secretary. The Act imposes additional requirements on the Secretary and submitting tribe with regard to energy resource agreements, including periodic reviews by the Secretary to monitor the activities of Indian tribes related to the development of energy resources.[11] Nothing in the tribal energy resource agreements absolves the United States from carrying out its trust obligations to Indian tribes.

The Administrators of Federal Power Marketing Administrations, including the Bonneville Power Administration and Western Area Power Administration (WAPA), are required to encourage tribal energy development.[12] Within two years after the enactment of the Act, the Secretary of Energy must report to Congress on tribal use of federal power allocations and any barriers that may impede tribal

---

[10] 2005 Energy Policy Act § 2604.

[11] 2005 Energy Policy Act § 2604(e).

[12] 2005 Energy Policy Act § 2605.

access and how to overcome such barriers. An appropriation of $750,000 is authorized for this effort.

The Secretary of Energy, in coordination with the Secretaries of the Army and Interior, must undertake a study on the cost and feasibility of developing a demonstration project on blending wind energy generated by tribes and hydropower generated by the Army Corps of Engineers on the Missouri River to supply firm power to WAPA.[13] A report to Congress on the study is due within one year of the enactment of the Act. An authorization of $1 million is appropriated to carry out this section.

[4] Consultation With Indian Tribes

The Act requires the Secretaries of Energy and the Interior to consult with Indian tribes in carrying out the provisions of the Tribal Energy Development and Self-Determination Act of 2005.[14]

[5] Four Corners Transmission Line Project and Electrification

Under the Act, the Dine Power Authority, a Navajo Nation enterprise, is eligible to receive grants and other assistance to develop a transmission line from the Four Corners area to southern Nevada.[15]

[6] Energy Efficiency in Federally Assisted Housing

The Secretary of Housing and Urban Development is required to encourage energy conservation and efficiency in Indian housing.[16]

[13] 2005 Energy Policy Act § 2606.

[14] 2005 Energy Policy Act § 504.

[15] 2005 Energy Policy Act § 505.

[16] 2005 Energy Policy Act § 506.

§ 1.06 Title VI—Nuclear

[1] Scope[1]

Nearly a fifth of the nation's electricity is generated at nuclear power plants.[2] There are now 104 fully licensed nuclear power reactors in the United States,[3] spread among 31 of the states,[4] and their operators are subject to regulation by the Nuclear Regulatory Commission (NRC) as licensees.[5] Despite nuclear power's significant position in the national portfolio of electricity supply, though, there have been no projects for the construction of new nuclear generating units in the United States (or Canada) in the past three decades.[6] Title VI of the Act is designed to provide for expanded use of nuclear power, to encourage the development of "next generation" nuclear generating facilities, and to ensure the safe operation of nuclear facilities.[7]

[2] Amendments to Price-Anderson Act

In 1957, Congress enacted the Price-Anderson Act, as a new Section 170 of the Atomic Energy Act of 1954 (AEA).[8] The Price-Anderson

---

[1] Not addressed within the scope of this section are those provisions of the Act that expands the tax deduction for qualified nuclear decommission fund contributions and establish a new production tax credit for power produced from certain advanced nuclear plants. Those provisions are discussed in the section on Title XIII, Energy Policy Tax Incentives.

[2] Energy Information Administration, Electric Power Monthly, October 2005, at 1 ("Year-to-date through July 2005, 49.6 percent of the Nation's electric power was generated at coal-fired plants . . . . Nuclear plants contributed 19.3 percent of the total generation, while natural gas-fired plants generated 18.4 percent, and 2.8 percent was generated at petroleum-fired plants. Conventional hydroelectric power provided 7.3 percent of the total, while other renewables and other miscellaneous energy sources generated the remaining electric power.").

[3] Only 103 of the 104 licensed units are operational at present. The exception is Brown's Ferry 1, which has not operated since 1985. The plant's owner-operator, the Tennessee Valley Authority, has stated that it intends to restart the reactor in mid-2007. Energy Information Administration, Electric Power Monthly, October 2005, at 1.

[4] Energy Information Administration report, "Nuclear Power: Features," available on EIA website at www.eia.doe.gov/cneaf/nuclear/page/newnuc.html.

[5] *See generally* Atomic Energy Act, 42 U.S.C. §§ 2011–2259.

[6] Energy Information Administration report, "New Reactor Designs," available on EIA website at www.eia.doe.gov/cneaf/nuclear/page/analysis/nucenviss2.html#ftn2.

[7] *See generally* 2005 Energy Policy Act §§ 601–657.

[8] 42 U.S.C. §§ 2011–2259.

Act was intended as a means of ensuring that adequate funds would be available to the public for the payment of liability claims in the event of a nuclear incident.[9] The statute also was intended to eliminate liability-related deterrents to private sector participation in nuclear activities.[10] The Price-Anderson Act requires as a primary matter that nuclear licensees obtain the maximum amount of private sector liability insurance that they can obtain, but in the event that that amount of coverage is not sufficient to cover claims arising from an accident, licensees must contribute to a fund that provides a secondary level of coverage. The 2005 Energy Policy Act extends, from current sunset dates of 2003 and 2006 until 2025, the indemnification provisions of the Price-Anderson Act, thus providing for public compensation for nuclear-related incidents well into the future.[11] The Act also increases the amounts that licensees may be required to pay into the referenced secondary coverage fund, from the current level of $63 million to $95.8 million per incident, and likewise increases the per-year limits from $10 million to $15 million.[12] Per-incident indemnification available (above applicable private sector insurance amounts) under the Price-Anderson Act is set at a maximum of $10 billion[13] (or $500 million for incidents outside the United States)[14] subject to adjustment for inflation.[15] Civil penalty liability of contractors that are not-for-profit institutions, relating to violations of regulatory requirements, is to be capped at the amount of fees that the institution has received under the relevant contract.[16]

---

[9] U.S. Department of Energy, Office of General Counsel, "Charter: Price-Anderson Task Force," available on DOE Office of General Counsel website at www.gc.doe.gov/price-anderson/charter.pdf.

[10] U.S. Department of Energy, Office of General Counsel, "Charter: Price-Anderson Task Force," available on DOE Office of General Counsel website at www.gc.doe.gov/price-anderson/charter.pdf.

[11] 2005 Energy Policy Act § 602.

[12] 2005 Energy Policy Act § 603.

[13] 2005 Energy Policy Act § 604.

[14] 2005 Energy Policy Act § 605.

[15] 2005 Energy Policy Act § 607.

[16] 2005 Energy Policy Act § 610.

[3] NRC Scholarship and Fellowship Program

The Act institutes a Scholarship and Fellowship Program within the NRC.[17] The purpose of the scholarship program is to broaden opportunities for the study of science, engineering, or other fields that the NRC may designate as critical to the regulatory mission of the NRC.[18] The fellowship program has comparable goals, but it applies to individuals in a graduate or professional degree program.[19] Upon completion of the scholarship or fellowship period, recipients are to take a position of employment at the NRC and are to work in that position for a period of time equivalent to one to three times the length of the scholarship or fellowship period.[20]

[4] Limitation of Antitrust Review

The Act exempts from antitrust review license applications filed with the NRC subsequent to the date of enactment of the Act.[21]

[5] Decommissioning Pilot Program in Arkansas

The Act authorizes the Department of Energy (DOE) to expend $16 million to establish a decommissioning pilot program under which DOE will decommission and decontaminate the sodium-cooled fast breeder experimental test-site reactor located in northwest Arkansas, in accordance with the decommissioning activities discussed in a 1998 DOE report.[22]

[6] Whistleblower Protection

The Act extends whistleblower protections for both DOE and NRC activities, including contractors, and provides for *de novo* review of whistleblower complaints in federal district court in the event of prolonged inaction.[23]

---

[17] 2005 Energy Policy Act § 622.

[18] 2005 Energy Policy Act § 622.

[19] 2005 Energy Policy Act § 622.

[20] 2005 Energy Policy Act § 622.

[21] 2005 Energy Policy Act § 625.

[22] 2005 Energy Policy Act § 628.

[23] 2005 Energy Policy Act § 629.

[7] Medical Isotope Production

Section 630 of the Act amends Section 134 of the AEA to impose requirements on the export of Highly Enriched Uranium (HEU) for medical purposes, intended to ensure the safe, timely and secure use of such material.[24] Previously, the referenced AEA provision sometimes caused licensing delays because it did not distinguish the use of HEU to produce medical radioisotopes from other uses. This provision permits medical isotope production matters to proceed on a separate track, provided that security concerns are addressed adequately by agreement.[25] The provision also obligates the NRC to review the adequacy of physical protection requirements applicable to the transportation and storage of highly enriched uranium for medical isotope production or control of residual material after irradiation and extraction of medical isotopes.[26]

[8] Safe Disposal of Certain Low-Level Radioactive Waste

The Act requires the DOE to designate and report to the Congress about the DOE's selection of an entity within the DOE to have the responsibility of providing a facility for safely disposing of all greater-than-Class C low-level radioactive waste.[27] The DOE's report to Congress is to address an estimate of the cost and a proposed schedule to complete an environmental impact statement and record of decision for a permanent disposal facility for greater-than-Class C radioactive waste.[28]

[9] Prohibition on Nuclear Exports to Countries That Sponsor Terrorism

The Act amends the AEA to prohibit expressly the transfer of any nuclear materials, nuclear equipment or sensitive nuclear technology to any country whose government has been identified by the Secretary of State as engaged in state sponsorship of terrorist activities.[29]

---

[24] 2005 Energy Policy Act § 630.

[25] 2005 Energy Policy Act § 630.

[26] 2005 Energy Policy Act § 630.

[27] 2005 Energy Policy Act § 631(b).

[28] 2005 Energy Policy Act § 631(b).

[29] 2005 Energy Policy Act § 632.

Exempted from this prohibition are exports, reexports, transfers, or retransfers of radiation monitoring technologies, surveillance equipment, seals, cameras, tamper-indication devices, nuclear detectors, monitoring systems, or equipment necessary to safely store, transport, or remove hazardous materials, except to the extent that such materials are available for use in the design or construction of nuclear reactors or nuclear weapons.[30]

[10] Demonstration of Hydrogen Production at Existing Nuclear Plants

The Act authorizes the DOE to expend up to $100 million to establish two projects, at geographically and climatically different sites, to demonstrate the commercial production of hydrogen at existing nuclear power plants.[31]

[11] Prohibition on United States Assumption of Liability for Certain Foreign Incidents

The Act prohibits any official of the United States government from entering into any arrangement that would impose liability on the United States or otherwise require an indemnity by the United States for nuclear incidents occurring in connection with the design, construction, or operation of a production facility or utilization facility in any country whose government has been identified by the Secretary of State as engaged in state sponsorship of terrorist activities.[32]

[12] Standby Support for Certain Nuclear Plant Delay Situations

The Act establishes a loan and insurance program to allow the DOE to enter into standby support contracts that would, under certain circumstances, cover costs of delays in commencement of operations of a "advanced nuclear facility," defined as any nuclear facility the reactor design for which was approved by the NRC at least as recently as 1994, or a facility having a substantially similar design of comparable capacity even if it was not approved by that date.[33] Specified levels

---

[30] 2005 Energy Policy Act § 632.

[31] 2005 Energy Policy Act § 634.

[32] 2005 Energy Policy Act § 635.

[33] 2005 Energy Policy Act § 638.

of compensation, through loans and insurance coverage, are available for up to a total of six reactors of no more than three different reactor designs. Amounts may be paid under this provision if the failure of a reactor to become fully operational on schedule was attributable to the NRC's failure to comply with schedules for review and approval of inspections, tests, analyses, and acceptance criteria established under the combined license or the conduct of pre-operational hearings by the Commission for the advanced nuclear facility, or litigation that delays the commencement of full-power operations of the facility. Where the specified circumstances are present, the DOE would pay 100% of covered costs for the first two reactors that qualify. However, the DOE would not cover any cost incurred through the failure of the project sponsor to take a required action or through events within the sponsor's control. Covered costs include principle and interest on debt coverage, along with the difference in the fair market price of purchased power and contractual price of power from the plant, up to a total of $500 million. For the next four reactors that qualify, the DOE would pay only 50% of the covered costs, up to $250 million.[34]

[13] Next Generation Nuclear Plant Project

In an effort to enhance the development and usage of advanced nuclear power technology, the Congress included in the Act provisions establishing a "Next Generation Nuclear Plant Project" and designating the DOE's Idaho National Laboratory (INL) as the lead organization to conduct research and to develop a prototype plant that would generate both electricity and hydrogen.[35] The Act thus authorizes $1.25 billion for fiscal years 2006-2015 to fund such a plant. The INL is to collaborate with other national laboratories, institutions of higher education, research institutes and domestic and international researchers to develop advanced technologies necessary for the integrated production of both electricity and hydrogen. The Act establishes a September 2011 deadline for completion of the first phase of the project.[36] In that initial phase, the INL and other entities involved in the project are to complete their selection and validation of the best high-temperature technology and initial design work on combining

[34] 2005 Energy Policy Act § 638.

[35] 2005 Energy Policy Act §§ 641–645.

[36] 2005 Energy Policy Act § 645.

electricity generation and hydrogen production in a single prototype plant, along with the best combination of reactor and plant. The DOE's Nuclear Energy Research Advisory Committee (NERAC) has the responsibility of reviewing all program plans for the project and monitoring progress on an ongoing basis.[37] The NERAC also is to ensure that any important scientific, technical, safety, and program management issues that arise receive all due attention, both within the project team and by the Secretary of Energy. The program contemplates a design competition, as well, among no more than four teams for not more than two years.[38] The competition is to be used to develop detailed proposals for competitive evaluation, and for selection of the final design of the prototype nuclear reactor and the associated hydrogen plant. The target deadline for the second phase of the project—completion of the construction of the prototype reactor and hydrogen plant project—is September 30, 2021.[39]

[14] Nuclear Security

Not surprisingly, post-9/11 reforms of nuclear security requirements are addressed extensively in the Act. Section 651 amends the AEA by adding new sections to govern security response evaluations at the nation's nuclear facilities, to require rulemaking proceedings on the NRC's revision of its policy on "design basis threat" (DBT), which is the hypothetical threat level intended to represent the greatest reasonable threat against which a regulated private guard force should be expected to defend.[40]

[a] Security Response Evaluations

As noted, the Act directs the NRC to perform extensive new security evaluations, including armed force exercises.[41] Those "force-on-

---

[37] 2005 Energy Policy Act § 643.

[38] 2005 Energy Policy Act § 645.

[39] 2005 Energy Policy Act § 645.

[40] 2005 Energy Policy Act § 651. The Act also imposes new requirements for the fingerprinting of employees of and visitors to nuclear facilities, § 652, establishes a safe harbor for security-related use of a wide range of firearms for the protection of nuclear facilities, § 653, and subjects to anti-terrorist provisions those persons who engage in sabotage of nuclear facilities, fuel or other nuclear-related materials, § 655.

[41] 2005 Energy Policy Act § 651.

force" exercises are required, to the maximum extent practicable, to simulate security threats in accordance with the current DBT applicable to a given nuclear facility.[42] The NRC is permitted to temporarily suspend these new security evaluation procedures for a particular facility designated as being under a heightened threat level, provided that the NRC has determined that the procedures would compromise security at the facility.[43] The NRC must submit a classified report annually to the Congress on the results of each security response evaluation conducted, along with any relevant corrective action taken by the affected licensee.

[b] Design Basis Threat Rulemaking

The Act requires the NRC to initiate a rulemaking proceeding within 90 days of the date of enactment of the Act, to revise the NRC's policy and procedures on DBT. In that rulemaking proceeding, the NRC is expressly bound to consider a wide range of factors set forth in the Act by Congress: (1) the events of September 11, 2001; (2) an assessment of physical, cyber, biochemical, and other terrorist threats; (3) the potential for attack on facilities by multiple coordinated teams of a large number of individuals; (4) the potential for assistance in an attack from several persons employed at the facility; (5) the potential for suicide attacks; (6) the potential for water-based and air-based threats; (7) the potential use of explosive devices of considerable size and other modern weaponry; (8) the potential for attacks by persons with a sophisticated knowledge of facility operations; (9) the potential for fires, especially fires of long duration; (10) the potential for attacks on spent fuel shipments by multiple coordinated teams of a large number of individuals; (11) the adequacy of planning to protect the public health and safety at and around nuclear facilities, as appropriate, in the event of a terrorist attack against a nuclear facility; and (12) the potential for theft and diversion of nuclear materials from such facilities.[44] The NRC issued a notice of proposed rulemaking (NOPR) on November 2, 2005, in compliance with this requirement of the Act.[45]

---

[42] 2005 Energy Policy Act § 651.

[43] 2005 Energy Policy Act § 651.

[44] 2005 Energy Policy Act § 651.

[45] Design Basis Threat, NRC RIN 3150-AH60, 70 Fed. Reg. 67380 (Nov. 7, 2005). The comment date for this NOPR is January 23, 2006.

[15] Federal Security Coordinators

The Act requires the NRC to assign a Federal Security Coordinator, employed by the NRC, to each of the NRC's designated regions.[46] Each Federal Security Coordinator's responsibilities are to include (1) communicating with the NRC and other federal, state, and local authorities concerning threats, including threats against such classes of facilities as the Commission determines to be appropriate; (2) monitoring such classes of facilities as the Commission determines to be appropriate to ensure that they maintain security consistent with the security plan in accordance with the appropriate threat level; and (3) assisting in the coordination of security measures among the private security forces at such classes of facilities as the Commission determines to be appropriate and federal, state, and local authorities, as appropriate.[47]

[16] Backup Power for Emergency Notification Systems

The Act directs the NRC to modify its regulations within 18 months of the date of enactment of the Act to provide for backup power for licensed nuclear power plants located such that there is a permanent population in excess of 15 million people within a 50-mile radius of the power plant. Specifically, the NRC is to require that backup power be available for the emergency notification system of the power plant, including the emergency siren warning system, in the event that the alternating current supply within the 10-mile emergency planning zone of the power plant is lost.[48]

[17] Radiation Source Protection

The Act also takes steps to monitor and better control the treatment and movement of radiation sources, building upon the Code of Conduct on the Safety and Security of Radioactive Sources adopted by the International Atomic Energy Agency (IAEA) in 2003.[49] This provision of the Act represents a significant change from the status quo, in that it gives regulatory jurisdiction over accelerator-produced radioactive material to the NRC, and the regulatory regime prescribed

---

[46] 2005 Energy Policy Act § 651.

[47] 2005 Energy Policy Act § 651.

[48] 2005 Energy Policy Act § 651.

[49] 2005 Energy Policy Act § 651.

by the Act is more stringent than that of the IAEA's Code of Conduct.[50] Thus, for example, the NRC is obligated by the Act to issue regulations prohibiting a person from exporting a radiation source unless the NRC has determined that (1) the recipient of the radiation source is permitted to receive and possess the radiation source under the laws and regulations of the country of the recipient; (2) the recipient country has the appropriate technical and administrative capability, resources, and regulatory structure to ensure that the radiation source will be managed in a safe and secure manner; and (3) before the date on which the radiation source is shipped, a notification has been provided to and received from the recipient country as the NRC deems appropriate.[51] Comparable regulations will govern radiation source imports, along with the creation of a tracking system for radiation sources within the United States.[52] Because of the potential effect of this new regulatory regime on medical treatments and devices known as "radiopharmaceuticals," the Act obligates the NRC in all of its rulemaking activities under these provisions to take into account the effects of any contemplated rule on radiopharmaceutical patients and physicians.[53]

§ 1.07 Title VII—Vehicles and Fuels

[1] Scope

Over the years, Congress debated the role that renewable energy would play in displacing the country's heavy reliance on fossil fuel and its byproducts.[1] While the debate continues around certain topics, Congress took steps in Title VII to promote alternative fuels for both private and commercial vehicles and to promote advanced technologies that increase the efficient use of fossil fuels in vehicles.

---

[50] 2005 Energy Policy Act § 651.

[51] 2005 Energy Policy Act § 651.

[52] 2005 Energy Policy Act § 651.

[53] 2005 Energy Policy Act § 651.

[1] Congressional Research Service, CRS Issue Brief IB10143, *Energy Policy: Comprehensive Energy Legislation (H.R. 6) in the 109th Congress* at 11 (July 29, 2005).

[2] Existing Programs

[a] Use of Alternative Fuels by Duel-Fueled Vehicles

The Energy Policy and Conservation Act (EPCA) required the Department of Energy (DOE), in cooperation with other agencies, to ensure that the maximum number of vehicles acquired for federal government use are duel-fueled (conventional and alternative fuel). However, a congressionally-mandated study by the National Academy of Science found duel-fueled vehicles were rarely operated on anything but conventional gasoline. At the same time, the duel-fuel designation had allowed manufacturers of such vehicles to sell otherwise less efficient vehicles while remaining in compliance with the corporate average fuel economy standards for new passenger vehicles. In response to this lack of compliance, Section 701 of the Act requires that duel-fueled vehicles acquired pursuant to the EPCA's mandate be operated on alternative fuels unless the DOE waives this requirement upon a finding that the alternative fuels needed are not reasonably available or that their costs are unreasonably higher than gasoline.[2]

[b] Incremental Cost Allocation

Section 702 amends the Energy Policy Act of 1992 (1992 Act) to require the General Services Administration to spread the costs of alternative-fueled vehicles across the entire fleet of vehicles distributed by the agency.[3]

[c] Alternative Compliance and Flexibility

Section 703 of the Act amends the 1992 Act to allow states and alternative fuel providers to petition the DOE for a waiver from the Mandate for Alternative Fuel Providers (the "Mandate") in the 1992 Act. The DOE must grant the waiver if the DOE finds the petitioner adequately demonstrated that it is in compliance with all of EPA's regulations under the Clean Air Act, and that it achieved a reduction in petroleum consumption equivalent to (1) what would result from 100% cumulative compliance with the fuel use requirements of the Mandate or (2) in the case of states, to the annual consumption by the state of alternative fuels if all of the cumulative alternative fuel

[2] 2005 Energy Policy Act § 701.

[3] 2005 Energy Policy Act § 702.

vehicles of the state given credit under the Mandate were to use alternative fuels 100% of the time.[4]

[d] Review of Energy Policy Act of 1992 Programs

The DOE must conduct a study by February 8, 2006 as to the effectiveness of Titles III (Alternative Fuels—General), IV (Alternative Fuels—Non-Federal Programs), and V (Availability and Use of Replacement Fuels, Alternative Fuels, and Alternative Fueled Private Vehicles) of the 1992 Act.[5]

[e] Report Concerning Compliance With Alternative Fueled Vehicle Purchasing Requirements

All federal agencies subject to the Energy Policy Act of 1992 must report on their compliance with that Act's alternative fueled vehicle purchasing requirements for the federal government fleet by February 15, 2006.[6]

[f] Joint Flexible/Hybrid Vehicle Commercialization Initiative

Section 706 of the Act requires the DOE to establish a program to improve technologies for commercialization of a combination hybrid/flexible fuel vehicle or a plug-in hybrid/flexible fuel vehicle.[7] Towards that end, the DOE must provide grants that give preferences to proposals that achieve the greatest reduction in miles per gallon of petroleum fuel consumption, achieve greater than 250 miles per gallon of petroleum fuel consumption, and have the greatest potential of commercialization to the general public within five years.[8] In support of these efforts, Congress appropriated $3 million for 2006, $7 million for 2007, $10 million for 2008, and $20 million for 2009.[9]

---

[4] 2005 Energy Policy Act § 703(a).

[5] 2005 Energy Policy Act § 704(a)–(b).

[6] 2005 Energy Policy Act § 705.

[7] 2005 Energy Policy Act § 706(a).

[8] 2005 Energy Policy Act § 706(c).

[9] 2005 Energy Policy Act § 706(f).

[g] Emergency Exemption

Vehicles used in the repair of transmission lines and in the restoration of electricity service following power outages are excluded from the definition of "fleet" in the 1992 Act.[10] Thus, vehicles used for these purposes need not comply with the fuel requirements under the 1992 Act.

[3] Hybrid Vehicles, Advanced Vehicles, and Fuel Cell Buses

[a] Hybrid Vehicles and Advanced Diesel Vehicles

The DOE must accelerate efforts to improve battery and other rechargeable energy storage systems, power electronics, hybrid system integration, and other technologies for use in hybrid vehicles.[11] The DOE also must establish a program, including making grants, to encourage domestic production and sales of efficient hybrid and advanced diesel vehicles.[12] Congress appropriated such sums as may be necessary to establish the program in 2006-2015.[13]

[b] Pilot Program for Advanced Vehicles

In consultation with the Department of Transportation (DOT), the DOE must establish a grant pilot program to be administered through the DOE's Clean Cities Program to provide grants to a maximum of 30 geographically dispersed projects to state and local governments as well as metropolitan transportation authorities.[14] The grants may be used to acquire alternative-fueled vehicles, hybrid vehicles, fuel cell vehicles, ultra-low sulfur diesel vehicles, and infrastructure supporting them.[15] The DOE must report to Congress the list of grant recipients and other applicants to the Pilot Program for advanced vehicles, and on the effectiveness of the program as a whole.[16] Congress authorized $200 million to carry out the Pilot Program.[17]

---

[10] 2005 Energy Policy Act § 707.

[11] 2005 Energy Policy Act § 711.

[12] 2005 Energy Policy Act § 712(a).

[13] 2005 Energy Policy Act § 712(b).

[14] 2005 Energy Policy Act § 721(a).

[15] 2005 Energy Policy Act § 721(b).

[16] 2005 Energy Policy Act § 722(a).

[17] 2005 Energy Policy Act § 723.

[c] Fuel Cell Transit Bus Demonstration

The DOE must establish a fuel cell transit bus demonstration program, in consultation with the DOT, to demonstrate at least 25 fuel cell transit buses and their supporting infrastructure in five geographically dispersed locations.[18] Congress appropriated $10 million for each of fiscal years 2006-2010 for this demonstration.[19]

[4] Clean School Buses

[a] Clean School Bus Program

The EPA, in consultation with the DOE and other responsible agencies, must establish a program to award grants on a competitive basis to eligible recipients for the replacement or retrofit of certain existing school buses with clean fuel technology.[20] In the case of replacement, priority with respect to awarding grants will be given for replacement of buses manufactured before the 1977 model year. In the case of retrofitting, priority will be given for retrofitting buses manufactured in or after the 1991 model year.[21] Congress appropriated $55 million for each of fiscal years 2006 and 2007, and such sums as are necessary for each of fiscal years 2008-2010.[22]

[b] Diesel Truck Retrofit and Fleet Modernization Program

The EPA, in consultation with the DOE, must establish a program for awarding grants to public agencies and entities for fleet modernization programs including installation of retrofit technologies for diesel trucks.[23] Only states, local governments, and their instrumentalities are eligible for assistance under this program.[24] Furthermore, the EPA will give preference in awarding grants to proposals that will achieve the greatest reductions in nonmethane hydrocarbons, oxides of nitrogen, and/or particulate matter or involve the use of EPA or California

[18] 2005 Energy Policy Act § 731(a).

[19] 2005 Energy Policy Act § 731(c).

[20] 2005 Energy Policy Act § 741(b).

[21] 2005 Energy Policy Act § 741(b)(2).

[22] 2005 Energy Policy Act § 741(d).

[23] 2005 Energy Policy Act § 742(a).

[24] 2005 Energy Policy Act § 742(b).

Air Resource Board-verified emissions control retrofit technology on diesel trucks that operate only on ultra-low sulfur diesel fuel after September 2006.[25] To support the program, Congress appropriated $20 million for 2006, $35 million for 2007, $45 million for 2008, and sums as are necessary for each of fiscal years 2009 and 2010.[26]

[c] Fuel Cell School Buses

The DOE must establish a program to enter into cooperative agreements with private sector fuel cell bus developers for the development and demonstration of fuel cell-powered school buses.[27] The non-federal funding for this program must be more than 20% for fuel infrastructure development and 50% for demonstration activities.[28] Congress appropriated $25 million for the 2006-2009 period to carry out this program.[29]

[5] Miscellaneous

[a] Railroad Efficiency

The DOE, in cooperation with the DOT and the EPA, must establish a cost-shared, public-private research partnership involving the federal government, railroad carriers, locomotive manufacturers and equipment suppliers, and the Association of American Railroads, to develop and demonstrate railroad locomotive technologies that increase fuel economy, reduce emissions, and lower costs of operations.[30] Congress appropriated $15 million for 2006, $20 million for 2007, and $30 million for 2008 to carry out this task.[31]

[b] Mobile Emission Reductions Trading and Crediting

By February 2006, the EPA must submit a report to Congress detailing its experience with the trading of mobile source emission reduction credits for use by owners and operators of stationary

[25] 2005 Energy Policy Act § 742(c)(2).

[26] 2005 Energy Policy Act § 742(f).

[27] 2005 Energy Policy Act § 743(a).

[28] 2005 Energy Policy Act § 743(b).

[29] 2005 Energy Policy Act § 743(d).

[30] 2005 Energy Policy Act § 751(a).

[31] 2005 Energy Policy Act § 751(b).

emission sources to meet emission offset requirements within a nonattainment area.[32]

[c] Aviation Fuel Conversion

By October 8, 2005, the Federal Aviation Administration and the EPA must study and identify the impact of air emissions on air quality in nonattainment areas, ways to promote fuel conservation measures for aviation to enhance fuel efficiency and reduce emissions, and opportunities to reduce air traffic inefficiencies that increase fuel burn and emissions.[33]

[d] Diesel Fueled Vehicles

The DOE must accelerate efforts to improve diesel combustion and after-treatment technologies for use in diesel-fueled vehicles.[34]

[e] Conserve by Bicycling Program

Section 755 requires the DOT to enter into a contract with the National Academy of Science to conduct and submit to Congress a study on the feasibility of converting motor vehicle trips to bicycle trips.[35] The DOT will also establish a pilot program to use education and marketing to convert motor vehicle trips to bicycle trips and to partner with interested parties in at least two fields to maximize bicycle facility investment and to facilitate ongoing programs that are sustained by local resources.[36] Congress appropriated $5.15 million to carry out the pilot program, $300,000 to publicize the program's results, and $750,000 for the study.[37]

[f] Reduction of Energy Idling

By December 8, 2005, the EPA must commence a review to update its mobile source air emission models used under the Clean Air Act to determine whether the models accurately reflect the emissions resulting from long-duration idling of heavy-duty vehicles and other

---

32 2005 Energy Policy Act § 752(a).

33 2005 Energy Policy Act § 753(a)–(b).

34 2005 Energy Policy Act § 754(b)–(c).

35 2005 Energy Policy Act § 755(b).

36 2005 Energy Policy Act § 755(c).

37 2005 Energy Policy Act § 755(e).

vehicles and engines.[38] The EPA must also review the emission reductions achieved by the use of idling reduction technology. The EPA must complete this review and update process no later than February 8, 2006. Congress authorized (1) $19.5 million for 2006, $30 million for 2007, and $45 million for 2008 to reduce extended idling from heavy-duty vehicles, and (2) $10 million for 2006, $15 million for 2007, and $20 million for 2008 to reduce extended idling from locomotives.[39]

[g] Biodiesel Engine Testing Program

Not later than February 8, 2006, the DOE must initiate a partnership with diesel engine, diesel fuel injection system, and diesel vehicle manufacturers as well as diesel and biodiesel fuel providers to determine the impact of biodiesel from different sources on current and future emission control technologies.[40] Congress authorized $5 million for each of fiscal years 2006-2010 to support the partnership.[41]

[h] Ultra-Efficient Engine Technology for Aircraft

The DOE must enter into a cooperative agreement with the National Aeronautics and Space Administration to develop ultra-efficient engine technologies for aircraft.[42] Congress appropriated $50 million for each of fiscal years 2006-2010 for this purpose.[43]

[i] Fuel Economy Incentive Requirements

Section 759 of the Act mandates that, in order for any duel fuel automobile to be eligible to receive the fuel economy incentives authorized under 49 U.S.C. § 32906(a)–(b), the automobile must have a label on its fuel compartment notifying that the vehicle can be operated on an alternative fuel and on gasoline or diesel, with the form of alternative fuel stated on the notice.[44] This requirement applies to vehicles manufactured on or after September 1, 2006.

[38] 2005 Energy Policy Act § 756(b).

[39] 2005 Energy Policy Act § 756(b).

[40] 2005 Energy Policy Act § 757(a).

[41] 2005 Energy Policy Act § 757(d).

[42] 2005 Energy Policy Act § 758(a).

[43] 2005 Energy Policy Act § 758(c).

[44] 2005 Energy Policy Act § 759.

[6] Automobile Efficiency

[a] Authorization for Appropriations for Implementation and Enforcement of Fuel Standards

In addition to any other funds already authorized, Congress appropriated $3.5 million for each of fiscal years 2006-2010 to the National Highway Traffic Safety Administration (NHTSA) to carry out its obligations with respect to average fuel economy standards.[45]

[b] Extension of Maximum Fuel Economy Increase for Alternative Fueled Vehicles

Congress extended the manufacturing incentives for alternative or duel fueled vehicles authorized under 49 U.S.C. § 32901 *et seq.*[46]

[c] Study of the Feasibility and Effects of Reducing Use of Fuel for Automobiles

By October 8, 2005, the NHTSA must study the feasibility and effects of significantly reducing the amount of fuel consumed by automobiles.[47]

[d] Update Testing Procedures

The EPA will update or revise the adjustment factors in various sections under the Code of Federal Regulations to take into consideration higher speed limits, faster acceleration rates, variations in temperature, use of air conditioning, shorter city test cycle length, current reference fuels, and the use of other fuel-depleting features.[48]

[7] Federal and State Procurement

[a] Federal and State Procurement of Fuel Cell Vehicles and Hydrogen Energy Systems

Section 782 of the Act requires that, by January 1, 2010, federal agencies using a light- or heavy-duty vehicle fleet must lease or purchase fuel cell vehicles and hydrogen energy systems to meet any

---

[45] 2005 Energy Policy Act § 771.

[46] 2005 Energy Policy Act § 772(a)–(b).

[47] 2005 Energy Policy Act § 773(a).

[48] 2005 Energy Policy Act § 774.

applicable energy savings goal developed by the DOE and the Hydrogen and Fuel Cell Task Force established under the Act.[49] In addition, the DOE may establish a cooperative program with state agencies that manage vehicle fleets to encourage the purchase of fuel cell vehicles. For this cooperative program, Congress allocated $15 million for 2008, $25 million for 2009, $65 million for 2010, and such sums as necessary for the each of fiscal years 2011-2015.[50]

[b] Federal Procurement of Stationary, Portable, and Micro Fuel Cells

By January 1, 2006, federal agencies that use electrical power from stationary, portable, or microportable devices must lease or purchase stationary, portable, or micro fuel cells to meet any applicable energy savings goals described in Section 808 of the Act.[51] An agency is exempt from this requirement if the DOE determines that such agency cannot find reliable stationary, portable, or micro fuel cells to meet the agency's needs.[52] For this purpose, Congress appropriated $20 million for 2006, $50 million for 2007, $75 million for 2008, $100 million for 2009, $100 million for 2010, and such sums as necessary for fiscal years 2011-2015.[53]

[8] Diesel Emissions Reductions

[a] National Grant and Loan Program

Section 792 of the Act requires the EPA to use 70% of the funds appropriated under this subtitle each year to provide grants and low-cost revolving loans to eligible entities to achieve significant reductions in diesel fuel emissions.[54] The EPA must also provide, at least, 50% of the funds available for a fiscal year under this section to eligible entities for the benefit of public fleets. In addition, the EPA must allocate, at least, 90% of funds available under this section for the development and commercialization of emerging technologies.

[49] 2005 Energy Policy Act § 782(b).

[50] 2005 Energy Policy Act § 782(e).

[51] 2005 Energy Policy Act § 783(b).

[52] 2005 Energy Policy Act § 783(b).

[53] 2005 Energy Policy Act § 783(d).

[54] 2005 Energy Policy Act § 792(a).

[b] State Grant and Loan Programs

The EPA must use 30% of the funds appropriated under this subtitle each year to support grant and loans programs administered by states that are designed to achieve significant reductions in diesel fuel emissions.[55]

[c] Evaluation and Report

Section 794 requires the EPA to provide a biennial report to Congress, evaluating the implementation of programs under this subtitle not later than August 7, 2006.[56]

[d] Outreach and Incentives

The EPA must establish a program to inform stakeholders, such as equipment manufacturers, emission and pollution control technology manufacturers, community organizations, and state and local air quality control officials, of the benefits of certain technologies and to develop non-financial incentives to promote their use. The EPA will also develop appropriate guidance to provide credit to states for emission reductions created by the use of these technologies.[57]

[e] Effect of Subtitle

Nothing in this subtitle affects any authority granted under the Clean Air Act.[58]

[f] Authorization of Appropriations

For carrying out this subtitle, Congress appropriated $200 million for fiscal years 2007-2011.[59]

§ 1.08 Title VIII—Hydrogen

[1] Scope

Title VIII of the 2005 Act is an outgrowth of prior initiatives and legislation aimed at development of fuel cell technologies. In 1990,

---

55 2005 Energy Policy Act § 793(a).

56 2005 Energy Policy Act § 794(a).

57 2005 Energy Policy Act § 795(c).

58 2005 Energy Policy Act § 796.

59 2005 Energy Policy Act § 797.

Congress passed the Spark M. Matsunaga Hydrogen Research, Development, and Demonstration Act to encourage basic research and development related to hydrogen energy and fuel cell technology.[1] Title VIII of the Act is a complete substitute[2] for that 1990 statute and provides for similar demonstration development as authorized in the earlier statute. Congress's goal was to create a mature hydrogen economy in the transportation sector to thereby decrease dependency on foreign oil, eliminate most transportation emissions, and enhance national security.[3] The Act also seeks to facilitate a partnership among private industry, institutions of higher education, National Laboratories, and research institutions with regard to hydrogen energy and fuel cell technology.

In addition, Title VIII continues the President's 2003 Hydrogen Fuel and FreedomCAR Initiative[4] and also provides for a hydrogen and fuel cell task force, authorizes projects that use solar and wind power to produce hydrogen, and directs the Secretary of Energy to enable technology developed to be transferred to the private sector.

[2] Plan for Hydrogen Fuel Cell Program

Within six months of enactment of the of the Act, the Secretary of Energy must transmit to Congress a plan that coordinates the hydrogen and fuel cell programs required by the Act and any other projects within the Department of Energy (DOE) that directly relate

---

1 42 U.S.C. § 12401.

2 2005 Energy Policy Act § 801.

3 2005 Energy Policy Act § 802.

4 FreedomCAR (Cooperative Automotive Research) was introduced in 2002 and aimed to increase efficiency of the automotive fleet. The Hydrogen Initiative announced in 2003 was intended to complement the FreedomCAR program by making available $720 million for research and development and focused on cooperative research between the federal government, universities, and private industry. *See* Congressional Research Service, CRS Report RS21442, *Hydrogen and Fuel Cell Vehicle R&D: FreedomCAR and the President's Hydrogen Fuel Initiative* (Dec. 17, 2004). Critics of the original initiative suggested the hydrogen program was intended to forestall attempts to significantly raise vehicle CAFE standards by relieving the automotive industry of pursuing technological innovation and was concerned that hydrogen vehicles may not be feasible. However, adequate investment in all potential alternative energy sources was considered an important part of a comprehensive energy policy. *See* Congressional Research Service, CRS Issue Brief IB10143, *Energy Policy: Comprehensive Energy Legislation (H.R. 6) in the 109th Congress* (May 31, 2005).

to fuel cells or hydrogen. This coordinated plan must include an agenda and milestones for evaluation of progress in the development of fuel cells and hydrogen technologies for the next five years. The plan also must describe the entities that will carry out the activities described in Title VIII and the expected role of each. The plan must also identify the most significant hurdles, both technical and nontechnical, that stand in the way of achieving the goals above. The plan must also describe its implicit policy assumptions, including those that would affect the sources of hydrogen or marketability of hydrogen products.[5]

[3] Hydrogen Development Programs

The Secretary of Energy, in consultation with other federal agencies and the private sector, must initiate a research and development program to demonstrate and commercialize the use of hydrogen for transportation, utility, industrial, commercial, and residential applications.[6] For example, the program shall address production of hydrogen from diverse energy sources including fossil fuels, hydrogen-carrier fuels, renewable energy resources, and nuclear energy. The program shall also address the use of hydrogen for power generation, the safe delivery and storage of hydrogen, the development of safe, affordable, and efficient fuel cells, and the ability of domestic automobile manufacturers to manufacture competitive hybrid vehicle technologies.[7]

Section 805(d) of the Act provides that the Secretary of Energy shall support education and research conducted at higher education institutions that relate to hydrogen and fuel cells. Enhanced public education and research in fundamental sciences, application design, and systems concepts is to be supported by the Secretary as part of the planning process.[8]

[4] Hydrogen and Fuel Cell Technical Task Force

Within four months of enactment of the Act, the President is to establish an Interagency Taskforce (Task Force). The Task Force is

---

[5] 2005 Energy Policy Act § 804.

[6] 2005 Energy Policy Act § 805(b).

[7] 2005 Energy Policy Act § 805(e).

[8] This includes education and research relating to materials, subsystems, manufacturability, maintenance, and safety. 2005 Energy Policy Act § 805(d).

to plan for safe, economical, and environmentally sound hydrogen fuel infrastructure, fuel cell applications, distributed power generation, uniform codes, and other factors necessary to promote the hydrogen and fuel cell programs. The Task Force is to be chaired by the Secretary of Energy and shall include representatives from various government agencies.[9] The Task Force must foster the exchange of information and technology among industry, academia, and the government. The Task Force must also maintain an inventory and assessment of hydrogen-related technologies and integrate technical and other information. The Task Force must also promote the marketplace introduction of infrastructure for hydrogen vehicles and conduct an education program that provides information to potential end-users of hydrogen technologies.[10]

[5] Technical Advisory Committee

In Section 807 of the Act, Congress established the Hydrogen Technical and Fuel Cell Advisory Committee. The Committee is to be comprised of members of industry, academia, and governmental agencies and shall advise the Secretary of Energy on hydrogen and fuel cell programs and activities.[11] The Advisory Committee is to review and make recommendations on the implementation of the hydrogen and fuel cell programs, the safety, economical, and environmental consequences of the technologies, and the Secretary's Plan. The Secretary must consider the Advisory Committee's recommendations but need not adopt them. Every two years, the Secretary must report to Congress all recommendations of the Advisory Committee and whether the Secretary has acted on the recommendations.[12]

---

[9] 2005 Energy Policy Act § 806(a) (including membership from the Office of Science and Technology Policy within the Executive Office, the Department of Transportation, Department of Defense, Department of Commerce, Department of State, Environmental Protection Agency, and the National Aeronautics and Space Administration).

[10] 2005 Energy Policy Act § 806(b)(2).

[11] Members appointed by the Secretary are to represent domestic industry, academia, professional societies, government agencies, federal laboratories, previous advisory panels, and financial, environmental, and other appropriate organizations. 2005 Energy Policy Act § 807(b)(1).

[12] 2005 Energy Policy Act § 807(d).

[6] Demonstration Projects

The Secretary of Energy must fund a limited number of demonstration projects related to hydrogen and fuel cell technologies. The funding must be used to devise concepts that provide for the use of advanced composite vehicles that reduce energy requirements and to design local distribution systems that would facilitate use of hydrogen. The Secretary should also select projects that include vehicle demonstrations as well as portable and stationary demonstrations of hydrogen energy technology, and those that address interdependency of hydrogen applications and infrastructure.[13] The vehicle demonstrations must explore the uses of hydrogen blends and hybrid technology including the use of such technology in internal combustion powered vehicles.[14] In keeping with the goal of increased public education, the Secretary must also select projects that raise awareness of hydrogen and fuel cell technology among the public. Additionally, the Secretary is to fund projects that facilitate identification of optimum technology among alternatives and carry out demonstrations in national parks, island areas, and on Indian tribal land.[15] As part of the demonstration program, the Secretary of Energy also must provide grants to devise system design concepts that provide for the use of advanced composite vehicles to reduce the use of energy.[16] The grants for the use of advanced composite vehicles will also be used for programs that are managed through an enhanced FreedomCAR program within the DOE that encourages involvement in cost-shared projects.[17]

Section 808(c) of the Act requires that the Secretary, in consultation with the Task Force and the Advisory Committee, to identify new requirements that refine technological concepts, planning, and applications. The Secretary shall identify new concepts after 2008 for stationary and portable applications and after 2010 for vehicles.[18] During the second phase of demonstrations, the Secretary shall

---

[13] 2005 Energy Policy Act § 808(a).

[14] Including demonstrations involving light-duty, heavy-duty, fleet, specialty industrial, and farm vehicles. 2005 Energy Policy Act § 808(a)(10).

[15] 2005 Energy Policy Act § 808(a)(6)–(9).

[16] 2005 Energy Policy Act § 808(b)(1)(A).

[17] 2005 Energy Policy Act § 808(b)(1).

[18] 2005 Energy Policy Act § 808(c)(1).

redesign subsequent program work to incorporate those requirements.[19]

[7] Codes and Standards

Section 809 of the Act provides for the timely development of safety codes and standards for fuel cell vehicles, hydrogen energy systems, and stationary, portable, and micro fuel cells. The Secretary of Energy shall rely on the Task Force, other agencies, and private organizations to develop the codes and standards. Similar to other sections of the Title, the Secretary shall support efforts by organizations and agencies to educate the public with regard to these safety codes and standards and to share information among the agencies and organizations relating to best practices.[20]

[8] Reports

No later than two years after the enactment of the Act, and every three years thereafter, the Secretary of Energy must submit a report to Congress describing the activities that are being carried out under Title VIII of the Act, including measures taken to support transition to a commercialized hydrogen economy, any changes in strategy, and progress made to achieving the production goals[21] for hydrogen-powered vehicles and hydrogen fuel stations.[22]

Title VIII requires the Secretary to enter into agreements with the National Academy of Sciences (Academy) under which the Academy will review the progress of the DOE's programs. The Academy's review, which will occur every four years, will cover program priorities and technical milestones and evaluate the DOE's progress in achieving these milestones. Within 45 days of receiving the review, the Secretary must transmit it to Congress with a plan to implement the Academy's recommendations, or with reasons that such recommendations should not be implemented.[23]

---

[19] 2005 Energy Policy Act § 808(c)(2).

[20] 2005 Energy Policy Act § 809(b).

[21] Production goals are 100,000 vehicles by 2010 and 2.5 million vehicles by 2020. 2005 Energy Policy Act § 811(a)(4)(A)–(B).

[22] 2005 Energy Policy Act § 811(a)(1)–(4).

[23] 2005 Energy Policy Act § 811(b).

[9] Solar and Wind Technologies

In Section 812 of the Act, Congress sought to encourage the development of solar and wind technologies that produce hydrogen. The Secretary of Energy must prepare detailed "roadmaps" for development of solar and wind energy technologies that result in the production of hydrogen. The Secretary must submit these "roadmaps" to Congress within four months of enactment of the Act. The Secretary must also provide for the establishment of five projects for these two energy technologies in regionally and climatically diverse areas to demonstrate the production of hydrogen at solar and wind energy facilities.[24]

In support of the development of solar and wind energy programs, the Secretary is to enhance fellowship and faculty assistance programs, provide support for research, and encourage collaborative research among industry, national laboratories, and higher education institutions. Other supports that are meant to increase communication and interaction among entities include developing sabbatical programs, visiting scientist programs, and fellowships so that Laboratories and higher education institutions can share personnel.[25]

[10] Technology Transfer

Under Section 813 of the Act, the Secretary shall carry out programs that provide for the transfer of critical technologies to the private sector and that accelerate wider application of those technologies in a global market. These programs must also foster the exchange of generic, nonproprietary information. The Secretary must also assess technical and commercial viability of technologies relating to the production, distribution, storage, and use of hydrogen energy and fuel cells.[26]

§ 1.09 Title IX—Research and Development

[1] Scope

Title IX of the Act creates the Energy Research, Development, Demonstration, and Commercial Act of 2005. The new provisions require the Secretary of Energy to establish programs to promote

[24] 2005 Energy Policy Act § 812(a)–(b).

[25] 2005 Energy Policy Act § 812(c)–(d).

[26] 2005 Energy Policy Act § 813(1)–(4).

increased efficiency and diversity of energy supply, decrease dependence on foreign supply, improve energy security, and decrease the environmental impact of energy-related activities.[1] The Secretary must publish cost and performance goals with each annual budget submission for energy efficiency in buildings, energy-consuming industries, and vehicles; electric generation, transmission and storage; renewable energy technologies; fossil energy (including resource recovery and power generation); and nuclear energy.[2]

[2] Energy Efficiency

The Secretary must establish programs to improve energy efficiency and environmental performance of vehicles, including:

- hybrid and electric propulsion systems;
- plug-in hybrid systems;
- advanced combustion engines;
- weight and drag reduction technologies;
- whole-vehicle design optimization; and
- advanced drive trains.[3]

Programs to improve the energy efficiency of buildings include both new construction and retrofits, and the use of a "whole-buildings" approach. Other programs include those to improve the process efficiency of energy-intensive and waste-intensive industries, and advanced control devices for electric motors.

For fiscal years 2007-2009, Congress appropriated $2.6 billion for the energy efficiency programs under Subtitle A.[4] None of this money can be used for implementation of energy efficiency regulations, weatherization programs, state energy conservation programs, or federal energy management measures under the National Energy Conservation Policy Act.[5]

---

1 2005 Energy Policy Act § 902(a).

2 2005 Energy Policy Act § 902(b).

3 2005 Energy Policy Act § 911(a).

4 2005 Energy Policy Act § 911(b).

5 42 U.S.C. § 8251 *et seq.*

[a] Next Generation Lighting Initiative

Within 90 days of enactment of the Act, the Secretary must select industry participants to participate in a next generation lighting initiative to develop solid-state lighting technologies.[6] The Secretary will make competitively-awarded grants to researchers, industry (through an industry alliance established by the Secretary), laboratories and schools. Funding for the next generation lighting initiative is set at $50 million a year for the years 2010-2013. The Secretary may grant preference to industry alliance members in competitively-selected awards for R&D, demonstration and commercial application and for licenses from inventors.

[b] National Building Performance Initiative

The Director of the DOE Office of Science and Technology Policy, within 90 days of enactment, must establish a national building performance initiative. The initiative is an interagency group that will seek to integrate federal, state and private sector efforts to reduce costs of construction, operation and maintenance, and renovation of buildings.[7] Within one year of enactment of the Act, the group must submit to Congress a plan for carrying out the appropriate federal role in the initiative. The Director of the Office of Science and Technology Policy must establish an advisory committee to make recommendations on the role the private sector may play in the initiative.

The Secretary, within 120 days of enactment, must enter into an agreement with the National Institute of Building Sciences to assess current voluntary standards for high performance buildings, determine if additional research is needed and recommend steps to accelerate development of standards.[8] The Secretary is required to establish a grant and technical assistance program to support development of voluntary standards for high performance buildings.

Section 915 of the Act requires the Secretary to establish a program to explore secondary uses of batteries used in electric vehicles, if the Secretary determines that there are sufficient batteries to support the program.[9] Within 180 days of enactment of the Act, the Secretary

[6] 2005 Energy Policy Act § 912.

[7] 2005 Energy Policy Act § 913.

[8] 2005 Energy Policy Act § 914.

[9] 2005 Energy Policy Act § 915.

must solicit proposals to demonstrate secondary uses of batteries and, within 90 days after receipt of the proposals, select up to five proposals to receive financial assistance. Selection criteria include diversity of battery type, geographic and climatic diversity, and life-cycle environmental effects.

An Energy Efficiency Science Initiative will award grants for research relating to energy efficiency.[10] The Secretary must submit an annual report to Congress detailing the activities of the Energy Efficiency Science Initiative, including a description of the grant award process and an explanation of how the research involves energy efficiency.

Within 18 months of enactment, the Secretary must establish Advanced Energy Efficiency Technology Transfer Centers by making grants to non-profits, state and local governments or universities.[11] Each Center will operate a program to encourage advanced energy methods and technologies through education and outreach. The Secretary will establish an advisory committee to advise on the creation of the Centers. The advisory committee members must have expertise in advanced energy methods and technologies, and must include representatives from:

- state or local energy offices;
- energy professionals;
- trade or professional associations;
- architects, engineers, or construction;
- the research community; and
- nonprofit energy or environmental organizations.

[3] Distributed Energy and Electric Energy Systems

Congress appropriated $768 million for fiscal year 2007 through fiscal year 2009 for research and development to improve the efficiency and reliability of distributed energy resources and systems,

[10] 2005 Energy Policy Act § 916.

[11] 2005 Energy Policy Act § 917. Section 917(g) defines Advanced Energy Efficiency Technologies as "all methods and technologies that promote energy efficiency and conservation, including distributed generation technologies, and life-cycle analysis of energy use."

including micro-cogeneration, fuel cells, advanced energy delivery, microturbines, reciprocating engines, thermally activated technologies, combined heat and power systems, load reduction, enhancing grid components, high-temperature superconductors, and other infrastructure technologies. Within one year of enactment, the Secretary must submit to Congress a five-year plan developed with input from the various sectors, to guide the Secretary's activity under this provision. No later than two years after submission of the five-year plan, the Secretary must report its progress under this provision to Congress.[12]

[4] Renewable Energy

Congress appropriated $2.227 billion for fiscal year 2007 through fiscal year 2009 for research programs for solar, wind, geothermal, hydropower, ocean energy, combined use of renewables, cogeneration of hydrogen and electricity, kinetic hydro turbines, and bioenergy. In carrying out those renewable energy provisions, the Secretary, in consultation with the Secretary of Agriculture, must demonstrate the use of renewable technologies to aid in the delivery of electricity to rural and remote areas, including use of advanced wind power technology, biomass, and geothermal energy. The Act seeks to develop advanced biochemical and thermochemical conversion technologies to make fuel from feed-stocks that can compete financially with gasoline or diesel in internal combustion engines or fuel cell-powered vehicles and reduce dependence on fossil fuels. The Act also establishes a program to examine the feasibility of using hydrogen propulsion in light-weight vehicles.[13]

[5] Agricultural Biomass Research and Development Programs

Subtitle D of the Act amends the Biomass Research and Development Act of 2000, to promote the economic viability of biomass fuels and to assesses the potential of federal lands as feedstock resources for biomass fuels. The Act contains production incentives for cellulosic fuels, grants for small business bio-product marketing, bio-economy development, and demonstration of cellulosic biomass innovations.

[12] 2005 Energy Policy Act §§ 921–925.

[13] 2005 Energy Policy Act §§ 926–940.

The Secretary of Agriculture must report to Congress within one year on the economic potential of bio-based products.[14]

[6] Nuclear Energy

Subtitle E addresses programs involving research on civilian nuclear energy, including increasing nuclear power in the United States energy portfolio, and reducing nuclear proliferation. The Act appropriates $1.18 billion for nuclear energy R&D for core programs for the period 2007-2009.[15] Another $420 million is appropriated for the same period for the operation and maintenance of nuclear energy research infrastructure, including for fuel recycling, disposal of radioactive sources and educational fellowships. The Act establishes several nuclear energy research programs, including the Nuclear Energy Research Initiative, a Nuclear Energy Systems Support Program, the Nuclear Power 2010 Program, and the Generation IV Nuclear Systems Initiative.[16] The Act also establishes an Advanced Fuel Cycle Initiative under which the Director of the Office of Nuclear Energy, Science and Technology will study proliferation-resistant fuel recycling for spent nuclear fuel.[17]

The Act also establishes a program to support civilian nuclear science programs in universities.[18] It also establishes a program to study ways to increase the safety and security of nuclear facilities from deliberate attacks and natural occurrences, and directs the Secretary to submit to Congress by August 1, 2006, survey results concerning the industrial applications of large radioactive sources.[19]

[7] Fossil Energy

Congress appropriated $1.878 billion for fiscal years 2007-2009 for the Secretary of Energy to conduct R&D, demonstration and commercial application programs to improve the efficiency, use of, and environmental performance of fossil energy.[20] The Secretary of

[14] 2005 Energy Policy Act §§ 941–948.

[15] 2005 Energy Policy Act § 951(b).

[16] 2005 Energy Policy Act § 952.

[17] 2005 Energy Policy Act § 953.

[18] 2005 Energy Policy Act § 954.

[19] 2005 Energy Policy Act §§ 956–957.

[20] 2005 Energy Policy Act § 961.

Energy is also required to conduct a program to facilitate production and generation of coal-based power, including research into innovations for existing coal plants, gasification systems, synthetic gas from coal, carbon capture and sequestration, mining technologies and mercury removal. Within 120 days of enactment of the Act, the Secretary must publish for public comment draft cost and performance goals for the cost-competitive use of coal. No later than 180 days after enactment of the Act, and every four years after, the Secretary must submit a report to Congress detailing final cost and performance goals.

The Act establishes a 10-year carbon capture program, including research and development of absorption techniques and chemical processes to remove carbon dioxide from gas streams amenable to sequestration.[21] The Act also establishes a research and development program on coal mining technologies, that would maximize production while minimizing waste and adverse environmental impacts. Section 965 of the Act requires the Secretary to establish a program to advance the science and technology available to domestic petroleum producers to minimize the economic dislocation caused by the decline of domestic supplies of oil and gas resources. The Secretary of the Interior must report to Congress every two years on the latest estimate of oil and gas reserves in federal and state waters off of Louisiana, Texas, Alabama, and Mississippi.[22] The Act also establishes a Low-Volume Oil and Gas Reservoir research program and a Complex Well Technology Testing Facility at the Rocky Mountain Oilfield Testing Center.[23]

The Act amends the Methane Hydrate Research and Development Act of 2000[24] to establish a program to promote research into methane hydrate as a potential alternative to natural gas. A total of $155 million is allocated for fiscal years 2006-2010.[25] Within 90 days of enactment of the Act, the Secretary, in consultation with the Secretaries of Commerce, Defense, and Interior, and the Director of the National Science Foundation, is required to commence the methane hydrate program.

---

21 2005 Energy Policy Act § 963.

22 2005 Energy Policy Act §§ 964–965.

23 2005 Energy Policy Act §§ 966–967.

24 Pub. L. No. 106-193, 114 Stat. 234 (2000).

25 2005 Energy Policy Act § 968.

[8] Science

Subtitle G appropriates a total of $13.939 billion for fiscal years 2007-2009 for research and support for facilities, education and analysis of high energy physics, nuclear physics, biological and environmental, basic energy sciences, advanced computing, and fusion energy.[26] Programs will include support for facilities, education, outreach, information, analysis and coordination of activities. The Act establishes a Fusion Energy Sciences Program to ensure that the United States is competitive in providing fusion energy, and allows for United States participation in the international burning plasma fusion research project. It further establishes research programs concerning catalysis, hydrogen, solid state lighting, advanced scientific computing, systems biology, fission and fusion, water supplies, and a spallation neutron source facility.[27] The Secretary must construct and operate a rare isotope accelerator.[28] Construction is to commence no later than September 30, 2008, and will be funded up to $1.1 billion. The Act establishes an education pilot program to enhance scientific, technological, engineering and mathematical literacy, postdoctoral energy research fellowship and science and technology scholarship programs.[29]

[9] International Cooperation

Subtitle H establishes a program to promote cooperation with other nations in the Western hemisphere and Israel on energy issues, including increasing production, improving efficiency, and the development of supply technologies.[30]

[10] Research Administration and Operations

Subtitle I mandates cost sharing for R&D programs, with at least 20% of funds coming from non-federal sources, except for research that is of a basic or fundamental nature. For demonstration and commercial operations, not less than 50% of costs are to come from non-federal sources. The Act requires that awards of funds under the

---

26 2005 Energy Policy Act § 971.

27 2005 Energy Policy Act §§ 973–980.

28 2005 Energy Policy Act § 981.

29 2005 Energy Policy Act §§ 983–984.

30 2005 Energy Policy Act §§ 985–986(a).

Act must be competitive and require merit analysis.[31] The Secretary must establish advisory boards to conduct external technical reviews of Department programs.[32] The Secretary is required to report to Congress on a biennial basis on equal employment opportunity practices at National Laboratories and to develop a strategy for maintaining, developing, and building new facilities and infrastructure and closing unneeded facilities.[33] The Secretary must periodically analyze programs and develop a plan to coordinate among Department programs and submit the results to Congress within 12 months of enactment, and every four years thereafter.[34]

The EPA Administrator, in consultation with state and local officials in Michigan, is required to conduct a demonstration program in Southwestern Michigan on the effect of transported ozone and ozone precursors within two years of enactment.[35] The Secretary of Transportation, in consultation with the Department and the United States Arctic Research Commission, must provide annual grants of $3 million per year for fiscal years 2006-2011 to a university to establish and operate the Arctic Engineering Research Center in Fairbanks to improve construction of roads, bridges, structures, and other infrastructure in the Arctic region.[36] $61 million is appropriated for the Secretary of Commerce, in consultation with others, to establish the Barrow Geophysical Research Facility to support scientific research in the Arctic.[37]

[11] Ultra-Deepwater and Unconventional Natural Gas and Other Petroleum Resources

Subtitle J requires the Secretary to establish a R&D program for ultra-deepwater and unconventional natural gas and other petroleum resource exploration and production, including technology challenges facing small producers.[38] The Secretary is required to contract with

---

[31] 2005 Energy Policy Act § 989.

[32] 2005 Energy Policy Act § 990.

[33] 2005 Energy Policy Act §§ 992–993.

[34] 2005 Energy Policy Act § 994.

[35] 2005 Energy Policy Act § 996.

[36] 2005 Energy Policy Act § 997.

[37] 2005 Energy Policy Act § 998.

[38] 2005 Energy Policy Act § 999A.

a corporation organized as a consortium from industry and the scientific and educational communities to administer the program, including issuing awards for R&D activities.[39] The program is funded at $50 million per year from proceeds of oil and gas lease royalties for fiscal years 2007-2017, with and an additional $100 million per year for fiscal years 2007-2016.[40]

§ 1.10 Title X—Department of Energy Management

[1] Scope

In general, the purpose of Title X is to expedite the migration of new energy-related technologies into the market and to increase technology transfers from the Department of Energy (DOE) to the private sector. Title X also elevates certain staff positions at the DOE with responsibility for the Department's activities related to science and mathematics to promote studies in those areas. Further, Title X provides incentives for research and development activities that decrease the United States' dependence on foreign sources of oil and provides for prizes for breakthroughs in research, development and commercial application of new energy technologies.

[2] Improved Technology Transfer

One of the goals of the Act is to improve the transfer of technology among National Laboratories and single-purpose research facilities. To accomplish this objective, and to promote the commercial application of promising energy technologies, Title X directs the Secretary of Energy to establish a Technology Transfer Coordinator,[1] a Technology Transfer Working Group,[2] and a Technology Commercialization Fund.[3] The working group is to be made up of representatives of DOE's various National Laboratories (e.g., Lawrence Livermore National Laboratory, Los Alamos National Laboratory, Oak Ridge National Laboratory, etc.) and single-purpose research facilities, and is to coordinate technology transfer activities and coordinate the

---

[39] 2005 Energy Policy Act § 999B.

[40] 2005 Energy Policy Act § 999H.

[1] 2005 Energy Policy Act § 1001(a).

[2] 2005 Energy Policy Act § 1001(d).

[3] 2005 Energy Policy Act § 1001(e).

exchange of information about practices, including information regarding the resolution of disputes involving intellectual property rights.[4] The working group also is to develop and distribute information about opportunities and procedures for technology transfers.[5] Title X provides that 0.9% of the DOE's annual budget for research and development is to be used to provide matching funds with private partners to promote the commercialization of promising energy technologies.[6] The Secretary of Energy is to submit to Congress a technology transfer execution plan, and provide annual updates to the plan.[7]

[3] Technology Infrastructure Program

The Act seeks to stimulate the development of technology clusters and to promote the sharing of technological expertise. Technology clusters are concentrations of technology-related businesses (i.e., scientific or engineering research endeavors, manufacturing interests employing new technology, and technology-related services providers), institutions of higher education, and non-profit organizations that support each other through either formal or informal relationships.[8] To further the goal of developing these technology clusters and to promote the sharing of technological expertise, Title X directs the Secretary of Energy to establish a Technology Infrastructure Program.[9] The Technology Infrastructure Program is aimed at improving the ability of National Laboratories and single-purpose research facilities to use and benefit from commercial research and encouraging the exchange of scientific and technological expertise.[10] The Act provides funding for technology projects that meet certain criteria and further the Department's mission.[11] For a project to qualify for funding, one of the following types of entities must take part in the project: (1) a business; (2) an institution of higher education; (3) a

[4] 2005 Energy Policy Act § 1001(d)(1)–(2).

[5] 2005 Energy Policy Act § 1001(d)(3).

[6] 2005 Energy Policy Act § 1001(e).

[7] 2005 Energy Policy Act § 1001(g).

[8] 2005 Energy Policy Act § 1002(a)(2).

[9] 2005 Energy Policy Act § 1002(b).

[10] 2005 Energy Policy Act § 1002(c)(2)–(3).

[11] 2005 Energy Policy Act § 1002(e).

non-profit organization; or (4) an agency of a state, local or tribal government.[12] Further, the Secretary of Energy must competitively select projects for funding under the program and the cost sharing principles outlined in Section 988 of the Act apply to these projects.[13] Projects are limited to a duration of five years or less. Title X outlines the criteria the Secretary of Energy is to use to determine which projects are to receive funding.[14] The program is funded with $10 million each year for fiscal years 2006-2008.[15] The Secretary of Energy is required to report to Congress by July 1, 2008, regarding whether the program should be continued.[16]

[4] Small Business Advocacy Assistance

Through the Act, Congress sought to encourage the increased participation of small businesses, including socially and economically disadvantaged small business concerns, in research and technology activities. To this end, the Act requires all National Laboratories to designate a Small Business Advocate, and provides the Secretary of Energy with discretion to impose the same requirement on single-purpose research facilities.[17] The primary goal of the Advocates is to increase the participation of small businesses in procurement, collaborative research, technology licensing and technology transfer activities. The Act also establishes a small Business Assistance Program, and provides $5 million per year through fiscal year 2008, to assist small business concerns in becoming more effective and efficient and to provide general technical assistance to small businesses to improve their products and services.[18]

[5] Improved Coordination and Management of Civilian Science and Technology Programs

The Act amends the Department of Energy Organization Act[19] to establish the office of the Under Secretary of Science within the

[12] 2005 Energy Policy Act § 1002(e)(2).

[13] 2005 Energy Policy Act § 1002(e)(3)(A).

[14] 2005 Energy Policy Act § 1002(f).

[15] 2005 Energy Policy Act § 1002(i).

[16] 2005 Energy Policy Act § 1002(h).

[17] 2005 Energy Policy Act § 1003.

[18] 2005 Energy Policy Act § 1003.

[19] 42 U.S.C. § 7101 *et seq.*

DOE.[20] The aim of this office is to better coordinate research and development programs within the DOE. The Under Secretary of Science is to be appointed by the President with the advice and consent of the Senate and is to serve as science and technology advisor to the Secretary.[21] Included among the responsibilities assigned to the Under Secretary of Science are responsibility for monitoring research and development programs to ensure no duplication or gaps and advising the Secretary of Energy on the management of National Laboratories, as well as on education and training activities.[22] The Under Secretary of Science also is to advise the Secretary on the grants and other forms of financial assistance necessary for effective research activities. In addition, the Under Secretary of Science is to advise the Secretary on the development of a strategic framework for the DOE's research and development activities.[23]

Congress determined that there should be an Assistant Secretary to oversee the DOE's missions in nuclear energy, and, accordingly, the Act amends the DOE Organization Act[24] to add a seventh Assistant Secretary to oversee the Department's efforts in nuclear energy.[25]

[6] Transaction Authority

The Act amends the Department of Energy Organization Act[26] to authorize the Secretary of Energy to enter into competitively-selected transactions.[27] This authority is similar to the authority of the Secretary of Defense to enter into such transactions, and is designed to provide a vehicle by which the Secretary can accomplish the objectives of the DOE in situations where entering into a contract or

---

20 2005 Energy Policy Act § 1006 (amending DOE Organization Act, 42 U.S.C. § 7132(b)).

21 42 U.S.C. § 7132(b)(1).

22 2005 Energy Policy Act § 1006(a) (amending DOE Organization Act, 42 U.S.C. § 7132(b)).

23 42 U.S.C. § 7132(b)(4)(f).

24 42 U.S.C. § 7133(a).

25 2005 Energy Policy Act § 1006(b) (amending DOE Organization Act, 42 U.S.C. §§ 7132(d), 7133(a)).

26 42 U.S.C. § 7101 *et seq.*

27 2005 Energy Policy Act § 1007 (amending DOE Organization Act, 42 U.S.C. § 7256(g)).

other form of agreement or providing a grant is not feasible.[28] The Secretary must use competitive, merit-based selection procedures when exercising this authority.[29] As required by the Act, the Secretary has issued proposed regulations to implement this new authority.[30] The proposed regulations permit DOE to enter into Technology Investment Agreements (TIA).[31] A TIA is defined as either a type of cooperative agreement with flexible provisions tailored to accommodate the financial management, property management, and purchasing systems of commercial firms, or a transaction other than a grant or cooperative agreement if the intellectual property provisions vary from statutory requirements applicable to the DOE.[32] The proposed regulations describe the instances in which the DOE may use TIAs and limits the use of TIAs to situations in which a for-profit firm is involved in the commercial application of the results of the project. The proposed regulations give contracting officers at the DOE considerable latitude to negotiate provisions in areas that have been sources of concern for commercial firms, and also list substantive issues that must be addressed in the award document. Further, the proposed guidelines require that recipients are to provide, to the maximum extent practicable, at least one half of the funding for the project, in order to ensure that the recipient of the TIA has a vested interest in the outcome of the project. Also, the proposed regulations prohibit the use of a TIA if the recipient is to receive a fee or profit, and explain that TIAs are intended to provide assistance to projects that are of mutual interest to the government and the recipient of the TIA, not to provide a fee for a good or service as is the case in government procurement contracts. The scheduled effective date for the proposed regulations is March 15, 2006. The Act also requires the Comptroller to report to Congress regarding the use of this authority after one year.[33] The authority granted to the Secretary under this Section of the Act is to expire on September 30, 2010.[34]

---

[28] 42 U.S.C § 7256(g)(4)(B).

[29] 42 U.S.C. § 7256(g)(4)(A).

[30] Assistance Regulations, 70 Fed. Reg. 69,250 (Nov. 15, 2005).

[31] Assistance Regulations, 70 Fed. Reg. 69,250 (Nov. 15, 2005), adding new Part 603 to 10 C.F.R.

[32] Assistance Regulations, 70 Fed. Reg. 69,250, 69,251 (Nov. 15, 2005).

[33] 42 U.S.C. § 7256(g)(6); 42 U.S.C. § 7256(g)(8)(b).

[34] 42 U.S.C. § 7256(g)(10).

[7] Prizes for Achievements in Grand Challenges

The Act authorizes the Secretary of Energy to award cash prizes in recognition of breakthrough achievements in research, development and commercial application that further the mission of the DOE.[35] The prizes may be awarded for achievements in a specific area designated by the Secretary of Energy, and the Act also specifically authorizes prizes (in cooperation with the Freedom Prize Foundation) for breakthroughs in processes and technologies that have the effect of reducing the United States' dependence on foreign sources of oil.[36] The Act authorizes $5 million for prizes for breakthroughs that reduce United States dependence on foreign oil and $10 million for prizes in other areas.[37]

[8] Transfer of Coal Research and Development to Department of Energy

The Act amends the Coal Research and Development Act of 1960[38] to transfer from the Department of the Interior to the DOE the responsibility for conducting research into ways to improve mining, preparing, and utilizing coal.[39]

[9] University Collaboration

Title X requires the Secretary of Energy to submit a report to Congress within two years of enactment examining the feasibility of promoting collaborations between the DOE and major universities and colleges. This report also is to consider providing incentives to increase the participation of small institutions of higher education, including minority-serving institutions, in DOE grants, contracts and cooperative agreements.[40]

[10] Procurement and Inventory Control

The Act requires the Secretary of Energy to develop and implement more stringent procurement and inventory controls to prevent waste

---

[35] 2005 Energy Policy Act § 1008(a).

[36] 2005 Energy Policy Act § 1008(b)–(c).

[37] 2005 Energy Policy Act § 1008(e).

[38] 30 U.S.C. § 661 *et seq.*

[39] 2005 Energy Policy Act § 1009(a).

[40] 2005 Energy Policy Act § 1010.

and fraud and directs the DOE's Inspector General to continue to closely review procurement and inventory practices.[41]

§ 1.11 Title XI—Personnel and Training

[1] Scope

To address concerns regarding the lack of qualified scientists in the energy field, Title XI of the Act creates various incentives for individuals to enter into energy-related research and development fields. This Title also creates technical training and continuing education programs for skilled workers in the energy field.

[2] Monitoring Workforce Trends

To keep abreast of trends in the energy workforce and to identify potential shortfalls in skilled personnel, the Secretary of Energy, with the assistance of the Secretary of Labor, is to monitor trends in the workforce of skilled technical personnel that support energy technology industries.[1] The Secretary of Energy is to report to Congress within one year on such trends and also is to make recommendations on how to meet future requirements for skilled personnel.[2] In addition, the Secretary of Energy is required to report to Congress as soon as practicable regarding any predicted significant shortage in skilled personnel in the energy technology industries.[3] Further, the Secretary of Energy, in consultation with the Secretary of Labor, is authorized to establish programs through which the Department will provide grants to enhance training for any workforce category for which the Secretary has identified a potential shortage.[4] The Act authorizes $20 million per year for fiscal years 2006-2008 to carry out these functions.[5]

[3] Education Programs in Science and Mathematics

To increase training and education in mathematics and the sciences, the Act amends the Department of Energy Science Education

[41] 2005 Energy Policy Act § 1011.

[1] 2005 Energy Policy Act § 1101(b)(1).

[2] 2005 Energy Policy Act § 1101(b)(2).

[3] 2005 Energy Policy Act § 1101(b)(3).

[4] 2005 Energy Policy Act § 1101(c).

[5] 2005 Energy Policy Act § 1101(d).

Enhancement Act[6] to require that at least 0.3% of Department of Energy (DOE) funds allocated to research and development be allocated to the Science Education Enhancement Fund to be used for training and education in these areas.[7] The funds are to be used to support competitive events designed to encourage student interest in and knowledge of science and mathematics, programs to promote professional development for mathematics and science teachers in grades from kindergarten through grade 12 at DOE research facilities, support summer internships at DOE research and development facilities, and to assist in educational and training activities for future workforce development in areas identified as critical skills needs.[8] The Act also authorizes the Secretary of Energy to provide funds to educational institutions to hire personnel to assist in coordinating educational efforts among local school systems, the DOE research and development institutions, and corporate and governmental entities.[9]

Title XI also directs the Secretary of Energy, in conjunction with the Secretary of Education, to conduct a study of the priorities, quality, flexibility and plans for educational programs at the DOE's research and development facilities. The study is to recommend measures to improve the DOE's coordination of its education, workforce development and critical skills development activities.[10] The Secretary of Energy must submit a report to Congress on the results of the study within two years of enactment.[11]

[4] Training Guidelines for Non-Nuclear Electric Energy Industry Personnel

The Act provides for the Secretary of Energy, in consultation with the Secretary of Labor, the electric industry and employee representatives, to develop model personnel training guidelines to support the

[6] 42 U.S.C. § 7381 *et seq.*

[7] 2005 Energy Policy Act § 1102(a) (amending DOE Science Education Enhancement Act, 42 U.S.C. § 7381a).

[8] 2005 Energy Policy Act § 1102(b) (amending DOE Science Education Enhancement Act, 42 U.S.C. § 7381b).

[9] 2005 Energy Policy Act § 1102(c) (amending DOE Science Education Enhancement Act, 42 U.S.C. § 7381c(b)).

[10] 2005 Energy Policy Act § 1102(e).

[11] 2005 Energy Policy Act § 1102(e)(3).

reliability and safety of the non-nuclear electric system. The guidelines are to include requirements for certification (and re-certification), competency and assessment for personnel in construction, operations, inspections and maintenance for transmission, generation, and distribution systems.[12]

[5] National Center for Energy Management and Building Technologies

The Act requires the Secretary of Energy to support and explore the possible expansion of the National Center for Energy Management and Building Technologies. The National Center is charged with the task of conducting research, education and training programs to improve the energy efficiency, indoor environmental quality, and security of both private sector and government buildings.[13]

[6] Improved Access to Energy-Related Scientific and Technical Careers

To improve the access of underrepresented groups to energy-related scientific and technical careers, Title XI requires the Secretary of Energy to give priority to programs that encourage the inclusion of these groups.[14] This section of the Act also amends the Department of Energy Science Education Enhancement Act[15] to require the Secretary of Energy to direct National Laboratories to increase the participation of historically Black colleges or universities, Hispanic-serving institutions, or tribal colleges in activities that further the goal of increasing the capacity of these institutions to train personnel in science or engineering disciplines.[16] These activities include collaborative research efforts, equipment transfers, training programs, and mentoring activities. The Secretary of Energy is directed to submit a report to Congress no later than two years after the enactment of the Act describing the activities carried out under this provision.[17]

---

12 2005 Energy Policy Act § 1103.

13 2005 Energy Policy Act § 1104.

14 2005 Energy Policy Act § 1105(a) (amending DOE Science Education Enhancement Act, 42 U.S.C. § 7381a(d)).

15 42 U.S.C. § 7381 *et seq.*

16 2005 Energy Policy Act § 1105(b) (amending DOE Science Education Enhancement Act, 42 U.S.C. § 7381 *et seq.*).

17 42 U.S.C. § 7381(d).

[7] National Power Plant Operations Technology and Educational Center

To address the training of certified operators and technicians of power plants, the Secretary of Energy is directed to support the establishment of a National Power Plant Operations Technology and Educational Center.[18] This Center is to be established at an institute of higher education that has expertise in providing training and offering degree programs in electric power generation, transmission and distribution technologies.[19] The Center is to provide training and continuing education, and must offer these programs both on-site and over the Internet.[20]

§ 1.12 Title XII—Electricity

[1] Scope

The 1992 Act provided a solid beginning for the creation of competitive interstate electric wholesale markets. Since that time, the utility industry has gone through a period of significant change and turmoil. In Title XII of the Act, Congress sought to further encourage the competition brought about by the 1992 Act, but also to rein in some of the competitive abuses that marred the industry in the early 2000s. To foster competition, Congress repealed the Public Utility Holding Company Act of 1935 (1935 Act), significantly amended Public Utility Regulatory Policies Act of 1978 (PURPA), and gave the Federal Energy Regulatory Commission (FERC) "backstop" authority to order the construction of transmission facilities if warranted by the national market. To rein in abusive behavior, Congress imposed mandatory reliability rules on the utility industry and its customers, established market transparency and anti-manipulation rules, and enhanced the FERC's authority to penalize entities that violate the Federal Power Act (FPA) or the FERC's orders, rules, and regulations. Many believe that the Act will provide the necessary regulatory platform for a more robust, but fair, wholesale electricity market to continue to develop.

---

[18] 2005 Energy Policy Act § 1106(a).

[19] 2005 Energy Policy Act § 1106(b).

[20] 2005 Energy Policy Act § 1106(c)(2).

[2] Electric Reliability Standards

On the afternoon of August 14, 2003, the northeast region of the United States and the Canadian province of Ontario experienced a power blackout. Nearly 50 million consumers were without electricity for several hours. After numerous investigations, a consensus emerged that one of the reasons for the blackout was that the rules of the North American Electric Reliability Council (NERC) that are designed to ensure the reliable operation of the nation's transmission grid were only voluntary. Each utility, although interconnected with others, could choose whether it would comply with the rules.

Reacting to the blackout, Congress gave the FERC jurisdiction over the reliability of the nation's bulk power system and authority to require that any reliability standard be mandatory. Thus, Title XII gives the FERC authority to establish an Electric Reliability Organization (ERO) to oversee and ensure the reliability of the nation's high-voltage bulk power grid.[1] Within 180 days of enactment of the Act, the FERC must issue new rules establishing the criteria an entity must meet to serve as the ERO, and the FERC must approve at least one ERO to act as the reliability entity for the nation's power grid.[2] An applicant seeking to be designated the ERO must show that it has the ability to develop and enforce reliability standards that provide for an adequate level of reliability of the bulk-power system. The applicant must demonstrate that it will be independent from the users, owners, and operators of the bulk-power system. The applicant also must show that it will equitably allocate dues, fees, and other charges among end users for all services provided; establish fair and impartial reliability standards and enforcement penalties; provide for reasonable notice and opportunity for public comment and a balance of interests and openness in developing reliability standards; and provide for obtaining recognition in Canada and Mexico.[3]

---

[1] 2005 Energy Policy Act § 1211. Section 1211 creates a new Section 215 in the FPA.

[2] 2005 Energy Policy Act § 1211 (b). On September 1, 2005, the FERC issued proposed rules in FERC Docket No. RM05-30-000 to comply with this section of the Act.

[3] 2005 Energy Policy Act § 1211(c). The NERC has announced that it is prepared to move quickly to become the ERO envisioned in the Act. Most in the industry believe the FERC will find that NERC qualifies as an ERO.

Once selected, the ERO will develop and file reliability standards with the FERC.[4] The FERC must approve the standards if the FERC finds them to be just and reasonable, not unduly discriminatory or preferential, and in the public interest. Regional entities organized on an Interconnection-wide basis may propose standards for their region, and the ERO must presume such proposals are just and reasonable.[5] In reviewing a proposed reliability standard, the FERC must give "due weight" to the technical expertise of the ERO and that of a regional entity. However, the FERC need not defer to the ERO or a regional entity with regard to standards addressing competition.[6] Further, the FERC will not be approving the new standards on a blank state. Therefore, the Act requires the FERC to create a procedure to address conflicts between a new reliability standard and any existing rules, orders, tariffs, or agreements imposed on an independent system operator (ISO), RTO or other regional transmission organization. Until the conflict is resolved by the FERC, the utility would continue to follow the preexisting order or rule.

The FERC may approve a reliability standard by rulemaking or by order.[7] If the FERC disagrees with a proposed standard, the FERC cannot revise the standard in the manner that it sees fit. Instead, the FERC must remand to the ERO any standard that does not meet the statutory requirements. However, the FERC may remand the standard with suggestions. Further, on it own motion or as the result of a complaint, the FERC may order the ERO to submit a proposed standard to address an issue if the FERC determines a new standard is necessary.[8] Thus, the FERC has some ability to direct the contents of a reliability standard. A reliability standard may not include any requirement to enlarge facilities, to construct new transmission or

[4] The Act defines a reliability standard to include a requirement to provide for the reliable operation of the bulk-power system, including requirements for the operation of existing bulk-power system facilities, including cybersecurity protection and the design of planned additions or modifications to facilities to provide for reliable operation of the bulk-power system.

[5] "Interconnection" is defined as a geographic area in which the operation of bulk-power system components is synchronized. 2005 Energy Policy Act § 1211(a).

[6] 2005 Energy Policy Act § 1211(d)(2).

[7] 2005 Energy Policy Act § 1211(d)(2).

[8] 2005 Energy Policy Act § 1211(d)(4)–(5).

generation capacity, or to set and enforce compliance with standards for the adequacy or safety of electric facilities or services.

The Act gives the FERC jurisdiction to enforce the reliability standards against all users, owners, and operators of the bulk-power system.[9] This includes the authority to require municipal organizations, and others typically exempt from FERC authority, to follow the ERO standards. Thus, the FERC may not only penalize a jurisdictional utility for failing to adhere to a standard, but the FERC may also penalize a traditionally non-jurisdictional transmission customer if it violates a reliability standard.

The Act gives the ERO first opportunity to impose penalties for violating a reliability standard. The ERO may impose a penalty after notice and an opportunity for a hearing. The ERO must then file the notice of the penalty and the record of the proceeding with the FERC. The penalty will take effect on the 31st day after the notice is filed. The penalized party may challenge the penalty by filing a request for review with the FERC within 30 days after the ERO files the notice of penalty. Filing a challenge does not stay the penalty, but the FERC may stay the penalty pending its review of the party's appeal. The FERC will then provide notice and an opportunity for a hearing with regard to a penalty imposed by the ERO. A party's hearing at the FERC may consist solely of a review of the record developed at the proceeding before the ERO and an opportunity to make a presentation as to whether the penalty should be adopted, revised or denied.[10] Thus, to ensure the FERC receives all relevant facts regarding the imposition of a penalty, a party should be sure to present all relevant evidence at the ERO phase of the process. After its review, the FERC may either affirm, set aside, reinstate or modify the penalty, or remand the case back to the ERO for further proceedings. In addition, the FERC may impose a penalty on its own motion after notice and an opportunity for a hearing. The Act makes clear that penalties must bear a reasonable relation to the seriousness of the violation and must take into consideration the violator's efforts to remedy the violation in a timely manner.[11] The Act requires the FERC to develop expedited review procedures for these penalty proceedings.

---

[9] 2005 Energy Policy Act § 1211(e).

[10] 2005 Energy Policy Act § 1211(e)(2).

[11] 2005 Energy Policy Act § 1211(e)(6).

The FERC is required to implement regulations that allow the ERO to delegate its authority for proposing and enforcing reliability standards to a regional entity, such as an ISO or RTO.[12] To receive the delegated authority, the regional entity must meet all of the criteria of an ERO. However, the regional entity may be governed by an independent board, a balanced stakeholder board, or a combination (independent and balanced) stakeholder board. Further, the delegation to the regional entity must promote efficient operation of the bulk-power system. The Act makes clear that the FERC and the ERO must rebuttably presume that a proposal for a delegation to a regional entity organized on an Interconnection-wide basis promotes efficient and effective operation of the power grid and should be approved. The FERC also may modify a delegation. Finally, the Act provides that the FERC may take action against the ERO or a regional entity to ensure compliance with a reliability standard or any order affecting the ERO or a regional entity.

The ERO must file any rules or rule changes with the FERC.[13] Any proposal to change an ERO rule must include an explanation of the proposed rule, the basis for the rule and its intended purpose. After notice and an opportunity for hearing, the FERC will accept the rule or change if it is just and reasonable, not unduly discriminatory or preferential, and in the public interest. The FERC or a party may also propose changes to the ERO's rules.

The Act requires the ERO to conduct periodic assessments of the reliability and adequacy of the bulk-power system in North America.[14] The Act provides that, after the President negotiates agreements with Mexico and Canada, the ERO will coordinate its reliability efforts with entities in those countries.[15]

The Act specifies that nothing in the reliability provisions of the Act is intended to preempt any state's authority to ensure the safety, adequacy, and reliability of electric service within the state, as long as the action is not inconsistent with any federal reliability standard.[16]

---

[12] 2005 Energy Policy Act § 1211(e)(4).

[13] 2005 Energy Policy Act § 1211(f).

[14] 2005 Energy Policy Act § 1211(g).

[15] 2005 Energy Policy Act § 1211(f).

[16] 2005 Energy Policy Act § 1211(i)(3). The Act provides that New York may adopt rules to establish greater reliability within New York, as long as the rules do not result in lesser reliability outside of New York.

The Act contemplates that the ERO or any other party may bring a conflict with state law to the FERC for review. Within 90 days of receipt of the filing, the FERC shall, after notice and an opportunity for a hearing, determine whether a state action is inconsistent with a reliability standard, "taking into consideration any recommendation by the ERO."[17] The FERC may, after consultation with the ERO and the state, stay the effectiveness of any state action pending the FERC's issuance of an order.[18]

The Act requires the FERC to establish a regional advisory body if asked by at least two-thirds of the states within a region that have more than one-half of their load served within the region.[19] The regional body will be composed of one member from each participating state in the region, appointed by the governor of each state. The body may also include representatives of agencies, states, and provinces outside the United States. The body may provide advice to the ERO, a regional entity, or the FERC regarding the governance of an existing or proposed regional entity within the same region, whether a proposed standard or penalty is just and reasonable and not unduly discriminatory and in the public interest, and any other issues requested by the FERC. The FERC may give deference to the advice of a regional body if the body is organized on an Interconnection-wide basis.[20]

[3] Transmission Infrastructure Modernization

The competition created by the 1992 Act changed the way the Nation's bulk power grid is used. As a result, the grid often experiences congestion that precludes or limits potential transactions. To foster new investment in transmission facilities that would alleviate that congestion, Congress gave the Department of Energy (DOE) the authority to identify regions where the bulk power grid is in need of additional transmission facilities. The expectation is that, once these areas are identified, transmission facilities would be installed to alleviate the congestion. Out of concern that states may delay or

---

[17] 2005 Energy Policy Act § 1211(i)(4).

[18] 2005 Energy Policy Act § 1211(i)(5).

[19] 2005 Energy Policy Act § 1211(j).

[20] The ERO provision does not apply to Alaska and Hawaii. 2005 Energy Policy Act § 1211(k).

prevent construction of transmission facilities in these regions, Congress gave the FERC "backstop" authority to issue permits for the construction of transmission facilities in those regions identified by DOE.[21]

[a] Siting of Interstate Electric Transmission Facilities and National Interest Electric Transmission Corridors

Within one year of enactment of the Act, and then every three years thereafter, the DOE, in consultation with affected states and regional entities, must perform a study of electric transmission congestion on the power grid. The purpose of the studies is to identify any geographic area experiencing transmission-capacity constraints or congestion that adversely affects consumers.[22] The regions where DOE finds such congestion will be designated a "national interest electric transmission corridor" (NIETC).[23] The Act gives the FERC "backstop" authority to authorize construction of transmission facilities in a NIETC over the objection of a state utility commission.

In determining whether to designate an area as a NIETC, the DOE must consider the following factors:

(1) Whether the economic vitality and development of the area or the end markets served in the area may be constrained by lack of reasonably priced electricity.

(2) Whether economic growth or the end markets in the area may be jeopardized by the limited sources of energy.

(3) Whether a diversification of supply is warranted.

(4) Whether the nation's energy independence would be served by the designation.

(5) Whether the designation would be in the interest of national energy policy.

(6) Whether the designation would enhance national defense and homeland security.[24]

---

21 2005 Energy Policy Act § 1221.

22 The permitting provisions of the Act do not apply in the Electric Reliability Council of Texas (ERCOT).

23 2005 Energy Policy Act § 1221; FPA § 216(a).

24 2005 Energy Policy Act § 1221; FPA § 216(a)(4).

Before the FERC can exercise its "backstop" permitting authority, it must first find that one of the following conditions exists:

(1) A state in which the facilities are to be constructed does not have the authority to approve the siting of facilities or consider the interstate benefits of such facilities.

(2) The applicant for the permit is a "transmitting utility" under the 2005 Act but does not qualify to file an application in the subject state because it does not serve end-use customers in the state.

(3) The state entity with authority to issue a permit withheld approval for more than one year after an applicant submitted an application or the designation of an area as a corridor (whichever is later).

(4) The state agency conditioned its approval in such a manner that the proposed construction will not significantly reduce transmission congestion in interstate commerce or is economically infeasible.

The FERC also must find that the proposed facilities will be used for the transmission of electric energy in interstate commerce, that the proposal is consistent with the public interest, that the proposal will significantly reduce congestion in interstate commerce and protect or benefit consumers, is consistent with sound national energy policy and will enhance energy independence, and will maximize, to the extent reasonable and economical, the transmission capabilities of existing towers and structures. The FERC must issue rules specifying the contents of a permit application.[25] The rules must allow affected states, governmental agencies, tribes, and property owners a reasonable opportunity to present their views on an application.[26]

The Act gives applicants the right to acquire the right-of-way over private property by exercise of the power of eminent domain in the U.S. District Court for the district in which the property is located, or in the appropriate court of the state in which the property is located.[27] Any right-of-way acquired under the Act must be used

[25] 2005 Energy Policy Act § 1221; FPA § 216(c).

[26] 2005 Energy Policy Act § 1221; FPA § 216(d).

[27] 2005 Energy Policy Act § 1221; FPA § 216(e).

exclusively for the construction or modification of electric transmission facilities and within a reasonable time after the acquisition. Any property obtained by eminent domain cannot be used for any other purpose, and the right-of-way will terminate if the authorized use terminates. Any right-of-way acquired under this provision is deemed to be a "taking," and the property owner must be compensated for the fair market value of the property taken.[28]

Likely in an effort to streamline the permit approval process, Congress designated the DOE to act as the lead agency for purposes of coordinating all federal authorizations and reviews of a proposed transmission facility.[29] The DOE must coordinate its efforts with any affected state, tribe, or multistate entity also involved in permitting and environmental reviews of a facility. The DOE, in consultation with other agencies, will establish deadlines and milestones for the necessary reviews of a proposed facility so that the DOE may ensure that all permit decisions and reviews required under federal law shall be completed within one year or as soon as practicable. The DOE must establish an expedited pre-application process to allow applicants to obtain information prior to filing an application that will indicate the likelihood of approval and key issues of concern to the agencies. The DOE will prepare a single environmental review report that will be used in all decisions regarding the proposed facility under federal law. If any agency has denied an application or fails to act by the required deadline established by the DOE, the applicant or any state in which the facility is located may appeal to the President. The President, in consultation with the affected agency, shall review the denial or failure to act. Within 90 days of the initiation of the appeal, the President may either grant the application with any appropriate conditions or deny the application.[30]

Within 18 months of enactment of the Act, the DOE must issue regulations to implement these provisions.[31] Within one year of enactment, the DOE and the heads of all federal agencies with authority to issue an authorization must enter into a Memorandum of

---

[28] 2005 Energy Policy Act § 1221; FPA § 216(f).

[29] 2005 Energy Policy Act § 1221; FPA § 216(h).

[30] In making a decision, the President is required to comply with federal law. 2005 Energy Policy Act § 1221; FPA § 216(h)(6)(D).

[31] 2005 Energy Policy Act § 1221; FPA § 216(h)(7)(A).

Understanding to ensure timely and coordinated reviews of applications. In exercising its responsibilities under this provision, the DOE is required to consult regularly with the FERC, RTOs, and other FERC-approved transmission entities.[32]

The Act authorizes three or more contiguous states to enter into a regional compact for the purpose of establishing a regional transmission-siting agency.[33] Such agencies will have the authority to review, certify, and permit siting of transmission facilities, including facilities in a NIETC. The FERC does not have "backstop" permitting authority in a region covered by a state agency, unless the members of the state agency cannot agree on a permit application.[34]

The Act makes it possible for third parties to contribute funds to upgrade transmission facilities owned by the Western Area Power Administration (WAPA) or the Southwestern Power Administration (SWPA).[35] The Act requires the FERC to encourage the deployment of advanced transmission system technologies, such as high-temperature lines, underground cables, advanced conductor technology, high-capacity ceramic wire, optimized line configuration, and others.[36]

[4] Transmission Operation Improvements

[a] Open Access by Non-Regulated Transmitting Utilities

Some have argued that the FERC's ability to mandate nation-wide open access transmission has been hindered because certain transmitting entities are not subject to the FERC's jurisdiction. In response to such claims, the Act adds a new Section 211A to the FPA authorizing the FERC to require unregulated transmitting utilities (*i.e.*, those otherwise excluded from FERC jurisdiction under Section 201(f) of the FPA, such as municipal and cooperative utilities) to provide open-access transmission services over their transmission facilities. These unregulated transmitting utilities must provide the open access transmission service at rates, terms and conditions that are comparable

32 2005 Energy Policy Act § 1221; FPA § 216(h)(9).

33 2005 Energy Policy Act § 1221; FPA § 216(i).

34 2005 Energy Policy Act § 1221; FPA § 216(i)(4).

35 2005 Energy Policy Act § 1222.

36 2005 Energy Policy Act §§ 1223, 1224.

to the rates, terms and conditions under which the unregulated entity provides transmission service to itself.[37] Section 211A of the FPA does not apply to the unregulated entities' distribution facilities.

An unregulated entity is exempt from the open access requirement if it sells less than four million megawatt-hours of electricity per year, does not own or operate any transmission facilities necessary for the operation of an interconnected system (or portion thereof), or meets other criteria that the FERC deems are warranted to exempt the transmission owner from having to provide open access service.[38] However, the FERC may terminate an exemption if it finds, after a hearing, that continuation of the exemption would unreasonably impair the continued reliability of the transmission grid.[39] In ordering the provision of open access transmission service, the FERC may not require an unregulated transmission entity to transfer control or operational control of its facilities to a Transmission Organization, such as an RTO, or to take action that would violate certain Internal Revenue Service rules regarding federal tax-exempt bond issuances.[40] Unregulated entities required to provide open access transmission service over their systems must file any rate, and any rate changes, with the FERC.[41]

[b] Federal Utility Participation in Transmission Organizations

Federal entities are authorized to transfer control and use of all or part of their transmission systems to transmission organization (RTO, ISO or other regional transmission entity authorized by the FERC).[42] The agreement under which the federal utility joins a transmission organization must include performance standards for operation and use of the transmission system to ensure recovery of all costs and expenses, to ensure adherence to existing contracts and financing arrangements, and to ensure compliance with statutory authorities, obligations, and limitations. The agreement must also include provisions for the federal

[37] 2005 Energy Policy Act § 1231; FPA § 211A(a)–(b).

[38] 2005 Energy Policy Act § 1231; FPA § 211A(c).

[39] 2005 Energy Policy Act § 1231; FPA § 211A(e).

[40] 2005 Energy Policy Act § 1231; FPA § 211A(i)–(j).

[41] 2005 Energy Policy Act § 1231; FPA § 211A(f).

[42] 2005 Energy Policy Act § 1232.

utility to monitor and oversee the transmission organization's terms and conditions, including a provision for resolution of disputes through arbitration and a provision to allow the federal utility to withdraw from the Transmission Organization.[43] Nothing in this provision, or actions taken pursuant to this provision, gives the FERC jurisdiction over the generation assets of a federal utility or over power sales by a federal utility.[44] Further, Congress made clear that implementation of this provision does not relieve a federal utility from adhering to other laws or contractual obligations.[45]

[c] Native Load Service Obligations

Since the issuance of Order No. 888, there has been tension between transmission owners and transmission users, and between the FERC and state utility commissions, over the extent to which transmission owners may reserve capacity on their transmission systems to serve native load customers. The Act adds Section 217 of the FPA to require the FERC to exercise its authority in a manner that facilitates the planning and expansion of transmission facilities to meet the reasonable native load obligations of load-serving entities (LSE). New Section 217 provides that LSEs may continue to use their firm transmission rights, or equivalent tradable or financial transmission rights, to meet their native load obligations.[46] To the extent that the service obligation covered by the firm transmission rights is transferred to another LSE, the successor LSE will also be entitled to use the firm transmission rights associated with the transferred obligation to serve native load.[47] New Section 217(k) to the FPA provides that any entity that exercises its rights over transmission to meet its native load obligation shall not be considered to engage in undue discrimination or preference under the FPA.

The Act does not affect any existing or future methodology employed by a Transmission Organization for allocating or auctioning transmission rights if the Transmission Organization was authorized by the FERC to allocate or auction financial transmission rights on

43 2005 Energy Policy Act § 1232(c).

44 2005 Energy Policy Act § 1232(d).

45 2005 Energy Policy Act § 1232 (e).

46 2005 Energy Policy Act § 1233(a); FPA § 217(b).

47 2005 Energy Policy Act § 1233(a); FPA § 217(b).

its system as of January 1, 2005, and the FERC determines that any future auction is just and reasonable and not unduly discriminatory or preferential.[48] If a Transmission Organization never allocated financial transmission rights prior to January 1, 2005, then, with respect to any application by the Transmission Organization to change its methodology, the FERC is required to exercise its authority in a manner consistent with the principles in the Act to protect transmission for native load use.

The Act authorizes the FERC to order LSEs to make transmission rights not used to meet a native-load obligation available to other entities in a manner that is just and reasonable and not unduly discriminatory or preferential.[49] The FERC may not, however, order a LSE to build transmission or distribution facilities.[50] The Act does not permit the FERC to abrogate any contract or service agreement for firm transmission service or rights in effect as of enactment of the Act.[51]

New FPA Section 217(g) directs the FERC to ensure that an unregulated entity that owns transmission facilities used predominately to support its own water pumping facility shall have protections for transmission service comparable to those provided to load serving entities. The Act also makes clear that a load serving entity located in the TVA and that has a firm wholesale supply contract with the TVA shall be considered to hold firm transmission rights for the transmission of the power provided. The native load provisions do not apply within the Electric Reliability Council of Texas (ERCOT).[52]

Within one year of enactment of the Act, the FERC must, by rule or order, issue new rules to implement in organized markets with transmission organizations the provision that the Commission shall act in a manner that facilitates the planning and expansion of transmission facilities to meet the reasonable needs of LSEs to serve their native load obligations.[53]

---

48 2005 Energy Policy Act § 1233(a); FPA § 217(c).

49 2005 Energy Policy Act § 1233(a); FPA § 217(d).

50 2005 Energy Policy Act § 1233(a); FPA § 217(e).

51 2005 Energy Policy Act § 1233(a); FPA § 217(f).

52 2005 Energy Policy Act § 1233(a); FPA § 217(h).

53 2005 Energy Policy Act § 1233(b).

[d] Study on the Benefits of Economic Dispatch

The DOE, in coordination with the states, must study the procedures currently used by electric utilities to perform economic dispatch.[54] The study is to identify possible revisions to economic dispatch procedures to improve the ability of nonutility or merchant generation to offer its output for sale via economic dispatch, as well as the potential benefits to residential, commercial, and industrial customers nationally and in each state if identified improved economic dispatch procedures were implemented.

The DOE must complete the first study within 90 days of enactment of the Act, and perform a new study each year thereafter. The report, which the DOE must send to Congress and the states, shall include any suggested legislative or regulatory changes.[55]

[e] Protection of Transmission Contracts in the Pacific Northeast

The Act creates new Section 218 of the FPA to provide for the protection of certain transmission rights for utilities that serve load in the Pacific Northwest. Under the Act, the FERC cannot require an entity that, as of August 8, 2005, held firm transmission rights on transmission facilities located in the Pacific Northwest (most of Oregon, Washington, Idaho, Montana, Wyoming, and Nevada) to convert its firm transmission rights to tradable or financial rights.[56] This provision is designed to ensure that, as the market in the Pacific Northwest evolves, current holders of certain firm transmission rights may retain those firm rights and not be required to convert them to another form of transmission rights.[57]

---

54 2005 Energy Policy Act § 1234. Economic dispatch is defined as the "operation of generation facilities to produce energy at the lowest cost to serve consumers reliably, recognizing any operational limits of generation and transmission facilities."

55 2005 Energy Policy Act § 1234(c). The DOE issued its report on November 21, 2005. The report called for more study to implement economic dispatch.

56 2005 Energy Policy Act § 1235.

57 16 U.S.C. § 218.

[f] Sense of Congress Regarding Locational Installed Capacity Mechanism

Responding to the controversy about the proposed adoption of a locational installed capacity (LICAP) mechanism in New England, the Act notes the concerns raised by the New England states and declares that the FERC should carefully consider the states' objections.[58] On August 10, 2005, the FERC issued an order delaying final action on the LICAP proposal so it could take oral argument and conduct further proceedings. The FERC indicated that the LICAP proposal would not be approved for implementation prior to October 1, 2006. On October 21, 2005, the Commission directed that a settlement judge be appointed to manage the settlement process, and required that any alternative developed in this process be submitted to the Commission by January 31, 2006.[59]

[5] Transmission Rate Reform

[a] Transmission Infrastructure Investment

As a result of Congress' desire to foster investment in the bulk power transmission grid, Section 1241 of the Act amends the FPA to add Section 219 to require the FERC to facilitate investment in transmission facilities. Within one year of enactment of the Act, the FERC must issue a rule establishing incentive-based (including performance-based) transmission rates.[60] The FERC's rule is to promote capital investment in the enlargement, improvement, operation, and maintenance of all facilities for the transmission of electric energy in interstate commerce, regardless of the ownership; provide a return on equity that attracts new investment in transmission facilities (including related transmission technologies); encourage deployment of transmission technologies and other measures to increase capacity and efficiency of existing transmission facilities and improve the operation of the facilities; and allow recovery of all prudently incurred investment necessary to comply with mandatory reliability standards

---

[58] 2005 Energy Policy Act § 1236.

[59] Devon Power LLC, 113 FERC ¶ 61,075 (2005).

[60] 2005 Energy Policy Act § 1241; FPA § 219(a). On November 18, 2005, FERC issued a proposed rule in Docket No. RM06-4-000 to comply with the Act's mandate. *Promoting Transmission Investment through Pricing Reform*, 113 FERC ¶ 61,182 (2005).

and all prudently incurred costs related to transmission infrastructure development.[61]

The FERC's rule also must provide incentives, to the extent within its jurisdiction, to each transmitting utility or electric utility that joins a Transmission Organization.[62] Further, the FERC must ensure that any costs recoverable pursuant to this section may be recovered by the utility through the transmission rates that it charges or the rates that the Transmission Organization charges to provide the service to the utility.[63] However, all rates approved under the FERC's new rules shall be subject to the requirements of Section 205 and 206 of the FPA, which require that all rates be just and reasonable and not unduly discriminatory or preferential.

[b] Participant Funding for New Interconnection and Transmission Upgrades

In an effort to facilitate construction of new generation, the FERC has made it easier for generators to connect their facilities to the grid by standardizing interconnection procedures and imposing certain interconnection costs on the Transmission Owner. Certain regional entities have sought to impose on generators (Participants) more of the costs of transmission upgrades. The Act authorizes the FERC to approve funding plans that allocate costs related to transmission upgrades or new generator interconnections, without regard to whether an applicant is a member of a FERC-approved Transmission Organization, to market participants. The funding plan must result in rates that are just and reasonable and not unduly discriminatory, and consistent with Sections 205 and 206 of the FPA.[64]

---

61 2005 Energy Policy Act § 1241; FPA § 219(b).

62 2005 Energy Policy Act § 1241; FPA § 219(c). A "transmitting utility" is any entity that owns or operates facilities for transmission of electric energy in interstate commerce, including entities such as municipalities otherwise exempt under Section 201(f) of the FPA.

63 2005 Energy Policy Act § 1241; FPA § 219(c).

64 2005 Energy Policy Act § 1242.

[6] Amendments to PURPA

[a] Net Metering and Additional Standards

The Act amends the PURPA to add three additional federal standards—net metering, fuel source diversity, and fossil fuel generation efficiency—that must be considered by state regulatory commissions and nonregulated electric utilities.[65] Within two years of enactment of the Act, each state regulatory commission and each nonregulated electric utility must commence consideration, or set a hearing date for consideration, regarding adoption of the new standards. This consideration must be completed within three years of enactment.[66] However, these timing requirements do not apply if, before the Act was enacted, the relevant state already had implemented a comparable standard, the state regulatory authority had conducted a proceeding to consider implementation of a comparable standard, or the state legislature had voted on the implementation of a comparable standard.[67] Finally, the Act amends Section 124 of the PURPA to make clear that the date of enactment of the provisions added by Section 1251 shall be the date of enactment of the Energy Policy Act of 2005.[68]

The Act defines net metering service to be service to an electric consumer under which electric energy generated by that electric consumer from an eligible on-site generating facility and delivered to local distribution facilities may be used to offset electric energy provided by the electric utility to the consumer during the billing period.[69] The Act defines fuel source diversity to require that each electric utility develop a plan to minimize dependence on one fuel source to ensure the utility can rely on a diverse range of fuel sources and technologies, including renewable technologies.[70] The Act defines fossil fuel generation efficiency to require each electric utility to develop and implement a 10-year plan to increase the efficiency of its fossil fuel generation.[71]

---

[65] 2005 Energy Policy Act § 1251(a); PURPA § 111(d)(11).

[66] 2005 Energy Policy Act § 1251(b)(1); PURPA § 112(b)(3)(A)–(B).

[67] 2005 Energy Policy Act § 1251(b)(3)(A); PURPA § 112(d).

[68] 2005 Energy Policy Act § 1251(b)(3)(B); PURPA § 124.

[69] 2005 Energy Policy Act § 1251(a); PURPA § 111(d)(11).

[70] 2005 Energy Policy Act § 1251(a); PURPA § 111(d)(12).

[71] 2005 Energy Policy Act § 1251(a); PURPA § 111(d)(13).

[b] Smart Metering

Within 18 months of the enactment of the Act, state regulatory commissions and nonregulated electric utilities must consider rules under which utilities will offer each of its customer classes, and provide to individual customers on request, a time-based rate schedule under which the rate charged for electric energy varies during different time periods and reflects the variance, if any, in the utility's costs of generating and purchasing electricity at the wholesale level.[72] The time-based rates should enable customers to manage energy use through advanced metering and communications technology. The time-based rates should include time-of-use pricing, critical peak pricing, real-time pricing, and credits for consumers with large loads that enter into load reduction agreements. However, the two year and three year implementation time requirements do not apply if, before the Act was enacted, the relevant state had implemented a comparable standard, the State regulatory authority had conducted a proceeding to consider implementation of a comparable standard, or the state legislature had voted on the implementation of a comparable standard.[73]

The Act requires a utility subject to Section 111 of the PURPA to provide a time-based meter to each customer that requests a time-based rate. In a state that allows third-party marketers to sell electric energy to retail electric consumers, those consumers must be entitled to the use of the same time-based metering and communications devices available to the utility's retail consumers.[74]

Congress designated the DOE to be responsible for promoting demand response initiatives.[75] Within 180 days of enactment of the Act, the DOE must submit a report to Congress that quantifies and identifies the national benefits of demand response and makes a recommendation for achieving specific levels of benefits by January 1, 2007. Within one year of enactment, the FERC must publish an annual report showing by region an assessment of demand response programs and steps taken to implement demand response initiatives.

---

[72] 2005 Energy Policy Act § 1252(a); PURPA § 111(d)(14).

[73] 2005 Energy Policy Act § 1252(i).

[74] 2005 Energy Policy Act § 1252 (a).

[75] 2005 Energy Policy Act § 1252(d); PURPA § 132(d).

[c] Cogeneration and Small Power Production Purchase and Sale Requirements

The Act changes the obligations that utilities have under PURPA to buy electric energy from, and to sell electric energy to, qualifying cogeneration and small power production facilities (QFs).

[i] Termination of Mandatory Purchases and Sale Requirements

Effective August 8, 2005, electric utilities are no longer required to enter into a new contract or obligation to purchase electric energy from a QF if such QF has non-discriminatory access to one of the following:

(1) Independently administered, auction-based day-ahead and real-time wholesale markets for the sale of electric energy and wholesale markets for long-term sales of capacity and electric energy;

(2) Transmission and interconnection services that are provided by a FERC-approved regional transmission entity and administered by an open-access transmission tariff that affords nondiscriminatory treatment to all customers and access to competitive wholesale markets that provide a meaningful opportunity to sell capacity to buyers other than the utility to which the QF is connected; or

(3) Wholesale markets for the sale of capacity and electric energy that are as competitive as those described under the first two conditions.[76]

An electric utility that meets one of these conditions may request approval from the FERC to be relieved of its purchase obligation on a service territory-wide basis.[77] The utility must set forth all facts that demonstrate that one of the three conditions have been met. After notice, including "sufficient notice" to potential affected QFs, the FERC must act on the request within 90 days of filing. Recognizing that markets may change over time, Congress provided that any QF, state agency, or other affected person subsequently may apply to the

[76] 2005 Energy Policy Act § 1253(a); PURPA § 210(m)(1).

[77] 2005 Energy Policy Act § 1253(a); PURPA § 210(m)(3).

FERC to have the purchase obligation reinstated because the above criteria are no longer met.[78]

As of late 2005, the FERC had addressed only two requests for relief of the purchase obligation. On August 25, 2005, Alliant Energy sought relief from the purchase obligation for new QFs in its region. On October 15, 2005, FERC dismissed Alliant Energy's request on procedural grounds.[79] The Commission explained that Section 1253 of the Act requires that all potentially affected QFs must receive "sufficient notice" of the application before the FERC may act on it. The FERC found that, although Alliant Energy's application was published in the Federal Register, Alliant Energy did not provide that required notice to "affected QFs." The Commission dismissed Alliant Energy's request without prejudice and provided the following guidance as to what the FERC expects a utility to show to satisfy the "sufficient notice" requirement. The FERC stated that it will not act on a request to terminate the purchase obligation unless, as part of the application, the utility identifies (including names and current addresses):

> (1) those QFs that have existing power purchase contracts with [the utility]; (2) other QFs that sell their output to [the utility] or that have pending requests for [the utility] to purchase their output; (3) any developer of generating facilities with whom [the utility has] agreed to enter into power purchase contracts or [is] in discussion with regard to power purchase contacts; (4) the developers of facilities that have pending state avoided cost proceedings; and (5) any other QFs that [the applicant] reasonably believes to be affected by its petition.[80]

The FERC issued a like ruling on October 17, 2005, in a proceeding involving Montana-Dakota Utilities Company.[81]

Under the Act, effective August 8, 2005, no utility will be required to enter into a new contract or obligation to sell electric energy to a QF if the FERC finds that competing retail electric suppliers are willing and able to sell and deliver electric energy to the QF, and the electric utility is not required by state law to sell electric energy in its service territory.[82]

---

[78] 2005 Energy Policy Act § 1253(a); PURPA § 210(m)(4).

[79] Alliant Energy Corporate Services, Inc., 113 FERC ¶ 61,024 (2005).

[80] Alliant Energy Corporate Services, Inc., 113 FERC ¶ 61,024, at P 20 (2005).

[81] Montana-Dakota Utilities Co., 113 FERC ¶ 61,045 (2005).

[82] 2005 Energy Policy Act § 1253(a); PURPA § 210(m)(5).

The elimination of QF purchase and sale obligations does not apply to any contract or obligation in effect, or that was pending approval before a state regulatory authority, on August 8, 2005.[83]

On January 19, 2006, the FERC issued a Notice of Proposed Rulemaking (NOPR)[84] to implement these new PURPA revisions providing for the termination of mandatory purchase and sale obligations under the Act.[85] The NOPR reflects the FERC's judgment that although the PURPA revisions are self-implementing in that they expressly permit electric utilities to file applications for relief from the mandatory purchase obligation, and require the FERC to act on such applications within 90 days, "the Commission has determined that it can more appropriately address this issue through rulemaking."[86] FERC Chairman Joseph Kelliher stated in an accompanying news release that the Commission's proposal "is based on a faithful reading of the statutory language, with the goal of providing regulatory certainty and minimizing administrative burdens."[87] Chairman Kelliher called the Act's scheme for revising utilities' purchase obligation under PURPA "a complicated one" and stated that the FERC's proposed rules "seek to provide as much clarity and certainty as possible given the statutory language."[88]

The January 19, 2006 NOPR sets forth the FERC's preliminary finding that utilities that are members of any of four named transmission entities (Midwest Independent System Operator, PJM Interconnection, ISO-New England, and New York Independent System

---

83 2005 Energy Policy Act § 1253(a); PURPA § 210(m)(6).

84 *New PURPA Section 210(m) Regulations Applicable to Small Power Production and Cogeneration Facilities*, 114 FERC ¶ 61,043 (2006), 71 Fed. Reg. 4532 (Jan. 27, 2006).

85 *See* 2005 Energy Policy Act § 1253; PURPA § 210(m).

86 *New PURPA Section 210(m) Regulations Applicable to Small Power Production and Cogeneration Facilities*, 114 FERC ¶ 61,043, at P 9 (2006), 71 Fed. Reg. 4532 (Jan. 27, 2006).

87 FERC News Release, "Pursuant to Energy Policy Act, Commission Proposes Revisions in PURPA Mandatory Purchase Obligation," Docket No. RM06-10-000 (Jan. 19, 2006), *available at* www.ferc.gov/press-room/press-releases/2006/2006-1/01-19-06-E-3.asp.

88 FERC News Release, "Pursuant to Energy Policy Act, Commission Proposes Revisions in PURPA Mandatory Purchase Obligation," Docket No. RM06-10-000 (Jan. 19, 2006), *available at* www.ferc.gov/press-room/press-releases/2006/2006-1/01-19-06-E-3.asp.

Operator) satisfy the Act's standards regarding independently administered real-time and day-ahead markets.[89] The NOPR indicates that such utilities therefore qualify for relief from the obligation to purchase power from QFs.[90]

As noted above, besides utility participation in sufficient day-ahead and real-time markets, the Act includes two other bases for utilities to shed their mandatory power purchase obligation—either (1) QF access to both a sufficient regional transmission entity and competitive wholesale markets,[91] or (2) QF access to competitively sufficient wholesale markets for the sale of capacity and energy.[92] As to these two categories, the FERC states in the NOPR that it will determine whether the criteria are satisfied on a case-by-case basis.[93] Applicants seeking relief from the PURPA purchase obligation under either of these two categories would be required, under the proposed rules, to provide, among other evidence, "actual sales data for (1) long-term and short-term capacity and (2) long-term, short-term, and real-time electric energy as well as evidence that the utility operates in a competitive wholesale market."[94]

The NOPR also sets forth proposed regulations reinforcing the new statutory provisions relieving utilities of any obligation to purchase power from new QFs.[95] Consistent with the *Alliant Energy* and *Montana-Dakota* proceedings discussed above, the NOPR sets for

---

89 *New PURPA Section 210(m) Regulations Applicable to Small Power Production and Cogeneration Facilities*, 114 FERC ¶ 61,043, at PP 12, 22–28 (2006), 71 Fed. Reg. 4532 (Jan. 27, 2006).

90 *New PURPA Section 210(m) Regulations Applicable to Small Power Production and Cogeneration Facilities*, 114 FERC ¶ 61,043, at PP 12, 22–28 (2006), 71 Fed. Reg. 4532 (Jan. 27, 2006).

91 PURPA § 210(m)(1)(B); *see* 2005 Energy Policy Act § 1253(a).

92 PURPA § 210(m)(1)(C); *see* 2005 Energy Policy Act § 1253(a).

93 *New PURPA Section 210(m) Regulations Applicable to Small Power Production and Cogeneration Facilities*, 114 FERC ¶ 61,043, at P 29 (2006), 71 Fed. Reg. 4532 (Jan. 27, 2006).

94 *New PURPA Section 210(m) Regulations Applicable to Small Power Production and Cogeneration Facilities*, 114 FERC ¶ 61,043, at P 29 (2006), 71 Fed. Reg. 4532 (Jan. 27, 2006).

95 *New PURPA Section 210(m) Regulations Applicable to Small Power Production and Cogeneration Facilities*, 114 FERC ¶ 61,043, at PP 33–35 (2006), 71 Fed. Reg. 4532 (Jan. 27, 2006).

proposed notice requirements for applicants for relief under any of the revised PURPA provisions.[96] Like the relevant sections of the Act, the NOPR provides a framework for the reinstatement in appropriate circumstances of the utility purchase obligation.[97]

Initial comments on the NOPR were scheduled to be filed by February 27, 2006, and reply comments were scheduled for March 28, 2006.[98]

[ii] Purchase and Sale Obligation for New Facilities

Upon enactment of the Act, an electric utility will no longer be required to buy electric energy from, or sell electric energy to, a new cogeneration facility unless the cogeneration facility can demonstrate to the FERC that the thermal energy output is used in a productive and beneficial manner and also that the electrical, chemical, and thermal output of the facility will be used fundamentally for industrial or commercial purposes rather than by an electric utility. On October 11, 2005, the FERC issued a proposed rule to establish these standards. Only new QFs that seek to obligate a utility to purchase its power will be subject to the new rules.[99] In other words, a facility that fails to meet the more stringent criteria cannot require a utility to buy its output.

[iii] Recovery of Avoided Costs

Under the PURPA, the rate the utility pays the QF for its output is often based on the utility's avoided cost. The avoided cost may be higher than the existing market rate. To ensure that utilities complying

[96] *New PURPA Section 210(m) Regulations Applicable to Small Power Production and Cogeneration Facilities*, 114 FERC ¶ 61,043, at PP 38–39 (2006), 71 Fed. Reg. 4532 (Jan. 27, 2006).

[97] *New PURPA Section 210(m) Regulations Applicable to Small Power Production and Cogeneration Facilities*, 114 FERC ¶ 61,043, at PP 42–44 (2006), 71 Fed. Reg. 4532 (Jan. 27, 2006).

[98] *New PURPA Section 210(m) Regulations Applicable to Small Power Production and Cogeneration Facilities*, 114 FERC ¶ 61,043, at P 58 (2006), 71 Fed. Reg. 4532 (Jan. 27, 2006).

[99] 2005 Energy Policy Act § 1253(a); PURPA § 210(m)(2). On October 11, 2005, the FERC issued a Notice of Proposed Rulemaking for Revised Regulations Governing Small Power Production and Cogeneration Facilities, FERC Docket No. RM05-36-000.

with the QF purchase obligation are able to recover the higher costs from its customers, the Act requires the FERC to issue new rules as necessary to ensure that an electric utility that purchases electric energy from a qualifying cogeneration or small power production facility in accordance with any legally enforceable obligation recovers all prudently incurred costs associated with the purchase.[100]

[iv] Ownership of QFs

Under the PURPA, Congress limited the participation of traditional electric utilities to owning up to 50% of a QF in an effort to foster new market entrants. Since passage of the PURPA in 1978, it became apparent that the ownership limit was no longer necessary. Therefore, the Act removes the ownership restrictions previously imposed on QFs by eliminating the prohibition on an electric utility owning more than 50% of the facility.[101]

[d] Interconnection

The Act amends the PURPA to require electric utilities to make available upon request "interconnection service" to any electric consumer that the electric utility serves.[102] "Interconnection service" means service to an electric consumer under which an on-site generating facility on the consumer's premises must be connected to the local distribution facilities. Within two years of enactment of the Act, state regulatory authorities and nonregulated utilities must adopt interconnection standards. This provision does not apply where a state regulatory agency adopted standards or commenced a proceeding to address standards prior to August 8, 2005.

[7] Repeal of the Public Utility Holding Company Act of 1935

The Act enacts a new law, the Public Utility Holding Company Act of 2005 (PUHCA 2005), that provides, among other things, for the repeal of the Public Utility Holding Company Act of 1935 (1935 Act). The repeal of the 1935 Act and enactment of the PUHCA 2005 are effective February 8, 2006.[103] As explained below, the new law, which

[100] 2005 Energy Policy Act § 1253(a); PURPA § 210(m)(7).

[101] 2005 Energy Policy Act § 1253(a); PURPA § 210(n).

[102] 2005 Energy Policy Act § 1254.

[103] 2005 Energy Policy Act §§ 1263, 1274.

contains some of the same recordkeeping and reporting requirements of the 1935 Act, will be implemented by the FERC. Thus, some have questioned whether it is accurate to state that the 1935 Act was *fully* repealed. The Act mandated the FERC to adopt rules implementing its provisions and on December 8, 2005, the FERC issued final rules to implement PUHCA 2005.[104] As the FERC noted in adopting these rules:

> [O]ur decisionmaking has been guided by the clear intent of Congress to repeal the regulatory regime established by PUHCA 1935 and to rely on state regulatory authorities and the Commission to protect energy customers, by supplementing the Commission's books and records authority under PUHCA 2005 and by enhancing our already significant authority over public utility mergers, acquisitions and dispositions of jurisdictional facilities. As we recognized in the notice of proposed rulemaking (NOPR), PUHCA 2005 is primarily a "books and records access" statute and does not give the Commission any new substantive authorities. In fact, the only substantive requirement contained in the new law is that we address requests involving certain allocations of costs of non-power goods and services. Accordingly, as discussed in greater detail below, we are rejecting requests that we re-impose particular requirements in PUHCA 1935 that Congress chose not to include in PUHCA 2005.[105]

[a] Background

When the Act was adopted, there were 29 public utility holding company systems registered, and subject to regulation, under the 1935 Act. The Securities and Exchange Commission (SEC) has regulated a broad array of transactions by registered holding companies and their subsidiaries. The 1935 Act has governed, among other things, the acquisition of securities (including securities of public utility companies) and utility assets, the issuance of securities, intercompany financings, and affiliate transactions.

The 1935 Act required that the operations of a holding-company system be limited to "a single integrated public-utility system, and to

[104] 2005 Energy Policy Act § 1272. Order No. 667, *Repeal of the Public Utility Holding Company Act of 1935 and Enactment of the Public Utility Holding Company Act of 2005*, 113 FERC ¶ 61,248, 2005 FERC LEXIS 2956 (2005) (Order 667), to be codified at 18 C.F.R. Part 366. In Order 667, the FERC announced its intention to hold a technical conference no later than February 8, 2007, to assess whether further changes to its rules may be required under PUHCA 2005.

[105] Order No. 667 at P 4.

such other businesses as are reasonably incidental, or economically necessary or appropriate to the operations of such integrated public-utility system." The requirement of an "integrated public utility system" has limited a holding company system to a single coordinated system confined in its operations to a single area or region. It has also limited the operations of a registered holding company system to energy-related businesses. Thus, these provisions of the 1935 Act have limited the ability of registered holding companies to engage in industrial or commercial enterprises. In addition, because of its restrictions, the 1935 Act also has limited the participation by others in the public utility industry. The requirement that the operations of the holding company and its subsidiaries be limited to energy-related activities has imposed an effective barrier to entry. Finally, the 1935 Act effectively limited the geographic spread of existing holding companies. Although the interpretation of a "single integrated public-utility system" has been expanded in recent years, it nonetheless has imposed an effective constraint on acquisitions by existing holding company systems. As a result of the 1935 Act, the public utility industry is generally fragmented, with utilities being confined to a limited geographic area.

The repeal of the 1935 Act will remove these geographic and business restrictions on holding companies. Besides the obvious relief from the requirement to obtain SEC approval for acquisitions of new utility assets and utility mergers as well as routine operational matters, such as financing and intercompany service agreements, the repeal of the 1935 Act removes restrictions on the types of investments—whether utility, energy-related, or otherwise—that can be made by holding companies. Further, the repeal of the 1935 Act removes constraints on companies in other businesses acquiring utilities and/or holding companies.

The repeal of the 1935 Act is expected to foster new investment in the Nation's bulk utility system. For example, geographically distant utilities will no longer be barred by federal law from merging and providing much needed investment in different regions. Further, it is expected that the repeal of the 1935 Act will mean that entities that did not want to have to deal with the regulatory burden of the 1935 Act may now invest in the utility industry. For example, because of the elimination of the geographical limits of the 1935 Act, expansion of multistate "transmission only" companies will now be easier and development of merchant transmission and investment in transmission by financial players and others may expand.[106]

[b] Federal Access to Books and Records

PUHCA 2005 authorizes the FERC to impose certain recordkeeping and reporting requirements on holding companies that, under the 1935 Act, were imposed by the SEC. Generally, PUHCA 2005 provides that each utility holding company and any affiliate thereof must provide to the FERC such books and records as the FERC determines are relevant to costs incurred by an electric public utility or natural gas company within such holding-company system and necessary or appropriate for the protection of utility customers with respect to FERC jurisdictional rates.[107]

The FERC rules in new Part 366 of its regulations (18 C.F.R.) implement PUHCA 2005. The text of Rule 366.2 largely repeats the statutory language—each "holding company," as defined in PUHCA 2005, must "maintain and make available" to the FERC its relevant books and records.[108] This provision, however, is not applicable to municipal or other governmental utilities.

Although the FERC had initially proposed adopting a number of the SEC's current reporting forms and other rules under PUHCA 1935, the final rules as adopted are more streamlined. All holding company systems, unless exempt or a waiver applies, as described below, must maintain records in sufficient detail to permit examination, audit and verification,[109] and must comply with the FERC's record retention rules.[110] Further, "service companies"[111] must keep their accounts in

---

[106] FERC has also granted a waiver from the holding company accounting, reporting, and records requirements to persons who invest in independent transmission only companies. *See* note 124 *infra.*

[107] 2005 Energy Policy Act § 1264.

[108] 18 C.F.R. § 366.2.

[109] 18 C.F.R. § 366.21(a).

[110] 18 C.F.R. § 366.21(b). The record retention rules are at 18 C.F.R. Parts 125 and 225.

[111] A service company is defined as any associate company within a holding company system organized specifically for the purpose of providing non-power goods or services or the sale of goods or construction work to any public utility in the same holding company system. 18 C.F.R. § 366.1.

accordance with the FERC Uniform System of Accounts[112] and must make an annual filing on new FERC Form No. 60.[113] This form will contain a detailed accounting of service company transactions with the utility companies in a system and is based on the SEC's Form U-13-60. However, the FERC has eliminated some of the requirements of the old form to comport with the more narrow focus of PUHCA 2005 versus PUHCA 1935. The FERC noted that this filing will help it in determining proper cost allocations among entities in a holding company system, will provide transparency, and will enable the FERC and others to better monitor for cross-subsidization.[114]

The foregoing recordkeeping, record-retention, notice filing, and accounting requirements will apply to all "holding companies." This is a broader definition than applied under PUHCA 1935; however, the FERC enacted a number of exemptions and waivers as discussed below.

All holding companies that meet the definition of PUHCA 2005 and FERC Rule 366.1 on February 8, 2006, whether or not exempt or entitled to a waiver, must make a notice filing with the FERC on Form FERC-65 by March 10, 2006; holding companies formed after February 8, 2006, must file on Form FERC-65 within 30 days after their formation.[115]

Under Rule 366.3, as required by PUHCA 2005, the FERC unconditionally exempts from all the accounting, record-retention, and reporting requirements of Part 366 of FERC's regulations any person that is a holding company solely by reason of holding one or more qualifying facilities (QFs), exempt wholesale generators (EWGs) or foreign utility companies (FUCOs).[116] In addition, the FERC

[112] 18 C.F.R. Parts 101 and 201. A transition rule allows currently registered holding company service companies to continue to use the SEC's uniform system of accounts for service companies during 2006.

[113] 18 C.F.R. § 366.23(a). The rules allow some transition period for companies that were exempt from similar requirements under PUHCA 1935. *See* 18 C.F.R. §§ 366.22(b)(2), 366.23(b).

[114] Order No. 667 at P 84.

[115] 18 C.F.R. § 366.4(a).

[116] 18 C.F.R. § 366.3(a). No notice to FERC is required in order to qualify for this exemption. The new rules also adopt a streamlined method for a person to claim the status of EWG or FUCO. *See* 18 C.F.R. § 366.7. EWG and FUCO have the same meanings as given in the 1935 Act.

exercised the discretion given it by Congress to provide an exemption from Part 366, upon application,[117] to the following classes of persons: passive investors including mutual funds, collective investment vehicles managed by banks, brokers/dealers, and other financial institutions; brokers/dealers, underwriters or fiduciaries who trade in public utility securities in the ordinary course of their business and do not exercise operational control over a utility; utilities subject to FERC jurisdiction that do not have captive customers; electric power cooperatives; and local distribution companies that are not regulated as natural gas companies.[118]

In addition to exempting persons who are technically holding companies but whose holdings do not give rise to a concern about affiliate transactions, as described in the preceding paragraph, the FERC exercised the discretion given it under the Act to exempt from its examination under PUHCA 2005, upon application by the holding company,[119] certain transactions. Transactions will not be subject to review where the holding company certifies that neither it nor its affiliates will charge, bill or allocate to the public utility or natural gas company in its system any costs or expenses in connection with goods and services transactions, and also will not engage in any financing transactions with such utility companies, except as authorized by a state commission or the FERC.[120] Further, upon application, the FERC will not review transactions between or among affiliates that are independent of and do not include a public utility or a natural gas company.[121]

The FERC also agreed in the new rules to grant blanket waivers of the accounting, record-retention and notification procedures, upon application,[122] to the following: single-state holding companies,[123]

[117] To obtain this exemption, the holding company must file a notice on new Form FERC-65A as prescribed in 18 C.F.R. § 366.4(b)(1). The person will be exempt immediately upon making a good faith filing. If the FERC does not take action on the filing within 60 days, the exemption is deemed to be automatically granted.

[118] 18 C.F.R. § 366.3(b)(1), (2), (5) and (6). The rule does not define "passive investor" but FERC indicated that it would follow its precedent for determining whether an investor is passive. Order No. 667 at P 120.

[119] *See* note 117 *supra*.

[120] 18 C.F.R. § 366.3(b)(3).

[121] 18 C.F.R. § 366.3(b)(4).

[122] To obtain this waiver, the holding company must file a notice on new Form

holding companies that control only 100MW or less of generation, and investors in transmission-only companies.[124]

Finally, the rules allow other persons to seek an exemption or waiver based on their specific circumstances if they can show that their books and records will not be relevant to the FERC's determination of jurisdictional rates.[125]

[c] State Access to Books and Records

PUHCA 2005 grants to state regulatory commissions authority to obtain access to books and records of a holding company and its affiliate companies if the state commission determines such books and records are relevant to costs incurred by the electric or gas distribution utility it regulates and access is necessary for the effective discharge by that state commission of its responsibilities.[126] However, the production of books and records shall be subject to necessary and appropriate safeguards against disclosure to the public of any trade secrets or sensitive commercial information. This provision does not preempt any state law or other law under which the state commission may otherwise have such access.[127]

[d] Affiliate Transactions

PUHCA 2005 confirms that nothing therein affects the authority of the FERC to require that jurisdictional rates are just and reasonable,

---

FERC-65B as prescribed in 18 C.F.R. § 366.4(c)(1). The person will be exempt (entitled to the waiver) immediately upon making a good faith filing. If the FERC does not take action on the filing within 60 days, the waiver is deemed to be automatically granted.

123 A "single-state holding company" is defined as one whose public utility operations are confined substantially to a single state. 18 C.F.R. § 366.1. Elsewhere, the FERC provides that if a holding company system does not derive more than 13% of its public-utility revenues from outside a single state, it will be deemed "confined substantially" to that single state. 18 C.F.R. § 366.5(b).

124 18 C.F.R. § 366.3(c).

125 18 C.F.R. § 366.3(d). To obtain such a waiver or exemption, the person must file a petition for a declaratory order pursuant to 18 C.F.R. § 385.207(a) justifying its request for exemption. 18 C.F.R. § 366.3(e).

126 2005 Energy Policy Act § 1265.

127 FERC did not adopt any new regulations implementing this portion of PUHCA 2005.

including the ability to deny or approve the pass-through of costs and the prevention of cross-subsidization between a utility and its affiliate.

[e] Applicability and Enforcement

Unless specifically provided, nothing in PUHCA 2005 applies to, or includes the United States, a state, any foreign governmental authority not operating in the United States or any agency or officer of the foregoing.[128] Further, nothing in PUHCA 2005 precludes the FERC or a state commission from exercising its jurisdiction under otherwise applicable law to protect utility consumers.[129] In addition, in Section 1270 of the Act, Congress made clear that the FERC shall have the same power as set forth in Sections 306 through 317 of the FPA to enforce the provisions of PUHCA 2005.[130]

[f] Savings Provision and Miscellaneous

Nothing in PUHCA 2005, or otherwise in the 1935 Act or rules, regulations or orders thereunder, prohibits a person from engaging in or continuing to engage in activities or transactions in which it was legally engaged or authorized to engage on the date of enactment of the Act, as long as that person continues to comply with the terms of any such authorization.[131] Further, nothing in PUHCA 2005 limits the authority of the FERC under the FPA or the NGA. Finally, tax treatment under Section 1081 of the Internal Revenue Code of 1986 as a result of transactions ordered in compliance with the 1935 Act shall not be affected in any manner due to the repeal of the 1935 Act and the enactment of PUHCA 2005.

In new Part 366 of its regulations, the FERC implements this savings provision by providing that any authorized activity may continue through the later of December 31, 2007 or the date of expiration of an SEC order giving such authorization under the 1935 Act.[132] In particular, Part 366 provides that any financing approval by the SEC under the 1935 Act remains effective, without the need for an equivalent FERC approval, if the holding company files the

---

128 2005 Energy Policy Act § 1268.

129 2005 Energy Policy Act § 1269.

130 2005 Energy Policy Act § 1270.

131 2005 Energy Policy Act § 1271.

132 18 C.F.R. § 366.6(a).

order with the FERC and complies with its terms (including filing with FERC the same filings and notices that were required to be made to the SEC).[133]

[g] Service Allocation

PUHCA 2005 also authorizes the FERC, at the election of a holding company system or an applicable state commission, to review and authorize the allocation of costs for non-power goods or administrative or management services provided by a service company to a public utility company in the same holding company system.[134] As required by PUHCA 2005, the FERC adopted a rule to exempt from these provisions any company in a holding-company system whose public utility operations are confined to a single state.[135] The FERC is authorized to exempt any other transactions that the FERC finds are not relevant to FERC jurisdictional rates of a public utility.[136]

On the issue of what price should be charged for a transaction between an affiliate and a public utility company, the FERC distinguished between traditional, centralized service companies and other non-regulated, affiliated special-purpose companies (such as a fuel or construction company). For service companies, the FERC indicated that pricing could be "at cost" as was the standard applicable under the 1935 Act. The FERC indicated that it would apply a rebuttable presumption that costs incurred under "at cost" pricing are reasonable. For the other service situations, the FERC indicated that when a utility provides service the price should be "no less than cost, i.e., the higher of cost or market." When the transaction is reversed, i.e., from a special purpose non-regulated affiliate to the utility, the price may not be above a market price.[137]

[8] Market Transparency, Enforcement, and Consumer Protection

The Act requires the FERC to facilitate price transparency in electric transmission and energy markets.[138] The market transparency and

[133] 18 C.F.R. § 366.6(b).

[134] 2005 Energy Policy Act § 1275.

[135] 18 C.F.R. § 366.5(b).

[136] 18 C.F.R. § 366.5(c).

[137] Order No. 667 at PP 169, 171.

[138] 2005 Energy Policy Act § 1281.

manipulation provisions, and the FERC's efforts to implement those provisions described in the "Oil and Gas" section of this chapter equally apply to the FERC's oversight of the electricity markets. *See* § 1.03 [Title III], *above.*

In addition, the Act makes clear that no entity (including unregulated entities exempt from the FPA under Section 201(f)) shall willfully and knowingly report to a federal agency any information related to the price of electricity sold at wholesale or the availability of transmission capacity, which the person or any other entity knew to be false at the time of reporting with the intent to fraudulently affect the data being compiled by a federal agency.[139]

[a] Penalties

The Act increases the criminal penalties for violations of the FPA from a maximum of $5,000 to a maximum of $1 million and the maximum jail time from two years to five years. The criminal penalties for violations of the FERC's rules, orders, regulations, restrictions, conditions, or order are increased from $500 per day of violation to $25,000 per day.[140] The FERC's authority to impose civil penalties is increased from a maximum of $10,000 to a maximum of $1 million. Also, the Act expands the civil penalty provisions to cover violations of any provision of the FPA, and not just certain sections (i.e., Sections 211 through 214), as was previously the case. Further, the Act amends the FPA to make clear that a court may prohibit, conditionally or otherwise, any individual who is engaged in or has engaged in providing false information from acting as an officer or director of an electric utility or from purchasing or selling electric energy or transmission services that are subject to the FERC's jurisdiction.[141]

[b] Refunds and Refund Effective Date

Prior to the Act, if the FERC found that rates were not just and reasonable, it could not order refunds of the unjust rates collected during the first 60 days following the institution of the complaint or proceeding.[142] The Act provides that a refund effective date may

139 2005 Energy Policy Act § 1282.

140 2005 Energy Policy Act § 1284.

141 2005 Energy Policy Act § 1288.

142 2005 Energy Policy Act § 1285.

commence upon the date a complaint or a proceeding is instituted rather than 60 days thereafter. In addition, the Act imposes refund obligations on municipal and other state agencies that are otherwise exempt from the FPA when those entities voluntarily make short-term sales (31 days or fewer, excluding monthly contracts subject to automatic renewal) through an organized market in which the rates are established by a FERC-approved tariff and the FERC finds that the sale violates the terms of the tariff.[143] However, if the entity sells less than eight million megawatt-hours of electricity per year or is an electric cooperative, it is exempt from the refund obligation. This provision is the result of the meltdown of the California market in 2000 and 2001 and will ensure that the FERC will have refund authority over all sellers in an organized market.

[c] Consumer Protection

The Act authorizes the Federal Trade Commission to issue rules protecting electric consumers from disclosure of personal information in connection with the sale or delivery of electricity.[144]

[9] Merger Review Reform

[a] Overview

The Act revises and expands the FERC's authority to review mergers, acquisitions, and dispositions of jurisdictional facilities under Section 203 of the FPA.[145] The major limit on the FERC's Section 203 authority is that Congress increased the dollar value of the facilities that triggers FERC review from $50,000 to $10 million. The Act expands the FERC's Section 203 authority because Congress has now clearly given the FERC jurisdiction over transactions involving only generation facilities.[146] Under prior law, the FERC did not have jurisdiction over a transaction involving only generating facilities. In addition, the FERC now has authority to review any transaction by which a holding company system that includes a transmitting utility

143 2005 Energy Policy Act § 1286.

144 2005 Energy Policy Act § 1287.

145 The amendments to Section 203 become effective six months after enactment. However, the new provisions do not apply to applications filed prior to enactment of the Act. 2005 Energy Policy Act § 1289(b)–(c).

146 2005 Energy Policy Act § 1289(a); FPA § 203(a)(1).

or an electric utility purchases or acquires any security with a value in excess of $10 million of, or merges or consolidates with, a transmitting utility, an electric utility, or a holding company system that includes a transmitting utility or an electric utility company with a value in excess of $10 million.[147] This provision, in conjunction with existing Section 204 of the FPA governing securities issuances, may give the FERC increased authority over intercompany lending transactions in holding company systems, such as money pool transactions.

The Act amends the FPA to add that, in evaluating applications under Section 203 of the FPA, the FERC must consider not only whether the transaction is consistent with the public interest, but also whether the transfer will result in cross-subsidization of a nonutility associate company or the pledge or encumbrance of utility assets for the benefit of any associate company in the holding company system.[148]

In the past, the FERC was criticized for taking too long to process Section 203 applications. As a result, Congress required the FERC to process certain Section 203 applications on an expedited basis. The FERC must classify transactions into categories that may easily meet the approval standards and then expedite review of such applications. For other transactions, the FERC must act on the application within 180 days of the date of filing. If the FERC does not act within 180 days, the application shall be deemed granted unless the FERC finds, based on good cause, that further consideration is required to determine whether the proposed transaction meets the public interest and does not result in a cross-subsidization. If the FERC finds more time is necessary, it may issue an order tolling the time to act on the application for an additional 180 days, at the end of which the FERC must grant or deny the application.[149]

[b] FERC's Implementing Regulations

On December 23, 2005, the FERC issued a Final Rule to amend its regulations to implement the changes to Section 203 of the FPA

[147] 2005 Energy Policy Act § 1289(a); FPA § 203(a)(2).

[148] 2005 Energy Policy Act § 1289(a); FPA § 203(a)(2).

[149] 2005 Energy Policy Act § 1289(a); FPA § 203(a)(5).

required by the Act.[150]

[i] Applicability

In Order No. 669, the FERC made clear that, under Section 203 as modified by the Act, the $10 million limit triggering FERC's jurisdiction does not apply to a public utility's merger or consolidation of its jurisdictional facilities with another, but instead only applies to a public utility's disposition of jurisdictional facilities, a public utility's acquisition of securities of public utilities, a public utility's purchase of existing generation facilities, and a holding company's acquisition of, or merger with, certain facilities. Thus, a public utility's proposal to "merge or consolidate, directly or indirectly, [jurisdictional facilities] or any part thereof with those of any other person, by any means whatsoever," will be subject to FERC's Section 203 jurisdiction regardless of the dollar value of the facilities to be merged or consolidated.[151]

[ii] Jurisdiction Over Acquisition or Purchase by a Holding Company in a Holding Company System and Blanket Waivers

The Act amends Section 203(a)(2) of the FPA[152] to require a holding company in a holding company system that includes a "transmitting utility" or an "electric utility" to obtain FERC approval prior to acquiring, purchasing or taking any security with a value in excess of $10 million of a "transmitting utility," an "electric utility company," or a holding company in a holding company system that includes a "transmitting utility" or an "electric utility." In addition, the Act requires a holding company in a holding company system that includes a "transmitting utility" or an "electric utility" to obtain FERC approval prior to merging or consolidating with a "transmitting utility," an "electric utility company," or another holding company system that includes a "transmitting utility" or an "electric utility" with a value in excess of $10 million.[153] Much of Order No. 669 focused

---

150 Order No. 669, *Transactions Subject to FPA Section 203*, to be codified at 18 C.F.R. §§ 2 and 33 (Order No. 669), *available at* www.ferc.gov/EventCalendar/Files/20051223121922-RM05-34-000.pdf.

151 Order No. 669 at P 32.

152 16 U.S.C. § 824b(a)(2).

153 2005 Energy Policy Act § 1289(a).

on how the FERC would apply Section 203(a)(2). The FERC ruled that "electric utility company," which Congress did not define, should have the same meaning as in PUHCA 2005: any company that owns or operates facilities used for the generation, transmission, or distribution of electric energy for sale.[154] Thus, the FERC's authority under Section 203(a)(2) is not limited to the acquisition of facilities used to sell or transmit power in interstate commerce. Indeed, this definition gives the FERC jurisdiction over an acquisition by a holding company of a distribution company, an entity that makes retail sales, and even a foreign utility company.

However, the FERC concluded that there would be no benefit in addressing on a case-by-case basis a holding company's purchase or acquisition of all types of "electric utility companies." Therefore, in Order No. 669 the FERC granted blanket authority to any holding company in a holding company system that includes a transmitting utility or an electric utility to purchase, acquire or take any security of:

(1) a transmitting utility or company that owns, operates, or controls only facilities used solely for transmission in intrastate commerce and/or sales of electric energy in intrastate commerce (i.e., within the Electric Reliability Council of Texas (ERCOT), Alaska, and Hawaii);

(2) a transmitting utility or company that owns, operates, or controls only facilities used solely for local distribution and/or sales of electric energy at retail regulated by a state commission; or

(3) a transmitting utility or company if the transaction involves an internal corporate reorganization that does not present cross-subsidization issues and does not involve a traditional public utility with captive customers.[155]

In addition, the FERC concluded that there are certain financial or other arrangements undertaken by holding companies that, while subject to the FERC's jurisdiction under Section 203(a)(2), do not harm competition or captive customers and, therefore, warrant blanket

[154] Order No. 669 at P 47; 18 C.F.R. § 33.1(b)(4).

[155] 18 C.F.R. § 33.1(c).

approval. These are transactions under which a holding company proposes to purchase, acquire, or take:

(1) any non-voting security (that does not convey sufficient veto rights over management actions so as to convey control) in a transmitting utility, an electric utility company, or a holding company that includes a transmitting utility or an electric utility company;

(2) any voting security in a transmitting utility, an electric utility company, or a holding company that includes a transmitting utility or an electric utility company if, after the acquisition, the holding company will own less than 10% of the outstanding voting securities; or

(3) any security of a subsidiary company within a holding company system.[156]

However, the blanket authorization granted to a holding company for these three transactions involving securities is subject to the holding company not borrowing from any electric utility company subsidiary in connection with the acquisition or not pledging or encumbering the assets of any electric utility company subsidiary in connection with the acquisition.[157] In addition, the FERC will require a holding company granted a blanket authorization for the three security transactions to submit to the FERC the same information that a holding company would have been required to submit to the SEC had the transactions occurred prior to the Act.[158]

In Order No. 669, the FERC also granted a conditional blanket authorization for holding companies in a holding company system that includes a transmitting utility or an electric utility to acquire foreign utility companies. The condition is that, if the holding company or any of its affiliates, subsidiaries, or associate companies within the holding company system have captive customers in the United States, the holding company must verify, through a duly authorized corporate official, that the transaction will not have an adverse impact on

156 18 C.F.R. § 33.1(c)(2).

157 18 C.F.R. § 33.1(c)(3).

158 18 C.F.R. § 33.1(c)(4). Thus, a purchaser would be required to file with the FERC SEC schedules 13D, 13G, and 13F. The FERC will publicly notice the filing of these schedules for informational purposes.

competition, rates, or regulation. Further, the verification must state that the transaction "will not result in [any of the following] at the time of the transaction or in the future":

(1) any transfer of facilities between a traditional utility associate company with wholesale or retail customers served under cost-based regulation and an associate company;

(2) any new issuance of securities by traditional utility associate companies with wholesale or retail customers served under cost-based regulation for the benefit of an associate company;

(3) any pledge or encumbrance of assets of a traditional utility associate company with wholesale or retail customers served under cost-based regulation for the benefit of an associate company; or

(4) any new affiliate contracts between non-utility associate companies and traditional utility associate companies with wholesale or retail customers served under cost-based regulation, other than non-power goods and services agreements subject to review by the FERC under Sections 205 and 206 of the FPA.[159]

A holding company's acquisition of a foreign utility company will be deemed approved upon the filing of the required information described above.[160]

Finally, the FERC held that Section 203(a)(2) does not apply to a transaction involving a holding company's acquisition of a governmental public utility, but clarified that Section 203 would apply to a non-governmental utility's acquisition of a non-governmental utility if the acquiring utility would plan to use the non-governmental utility's facilities to make wholesale sales in interstate commerce.[161] The FERC also ruled that Section 203(a)(2) applies to a holding company's

[159] 18 C.F.R. § 33.1(c)(5).

[160] 18 C.F.R. § 33.1(c)(5)(iii). The preamble to Order No. 669 also indicates that the FERC will grant blanket approval to routine cash management transactions that are designed to ensure no improper cross-subsidization, but this language was not codified. Order No. 669 at PP 142–143.

[161] Order No. 669 at P 58.

acquisition of a QF and an EWG.[162] Further, the FERC ruled that the term "holding company" includes an entity that owns or controls only a foreign utility company, EWG, or QF.[163] The FERC also concluded that, although the definition of holding company in the Act could be read to include governmental entities or electric power cooperatives, because those types of entities are exempt from FERC jurisdiction under Section 201(f) of the FPA, and the Act did not indicate any change to Section 201(f), the FERC would not read Section 203(a)(2) to apply to governmental entities or electric power cooperatives.[164]

[iii] Review of Generation-Only Transactions Clarified by FERC

Prior to the Act, the FERC did not have jurisdiction over a transaction that included the transfer or acquisition of only generation facilities.[165] To close that loophole, Congress gave FERC authority over the purchase, lease, or acquisition of an "existing generation facility" that has a value in excess of $10 million and is used for interstate wholesale sales subject to FERC's ratemaking jurisdiction.[166] In Order No. 669, the FERC defined "existing generation facility" as a generation facility that is operational at or before the time the transaction is consummated.[167] Further, the FERC ruled that "operational" means that the generation facility is complete, "*i.e.*, it is capable of producing power."[168] The FERC made clear that "operational" does not mean facilities that are only in the development or construction stage. However, the FERC also clarified that "operational" would include a facility that had been "mothballed," because

[162] The FERC explained that its regulations already exempt QFs from the filing requirements under Section 203, but that holding companies that own QFs are not exempt. Order No. 669 at P 59 n.54.

[163] The FERC noted that, although a holding company that solely owns QFs, EWGs, or foreign utility companies is exempt from PUHCA 2005, Congress did not include similar exemption language in the Act when addressing FERC's Section 203 jurisdiction. Order No. 669 at P 70.

[164] Order No. 669 at P 73.

[165] *E.g.*, Perryville Energy Partners, LLC on Behalf of Entergy Services, Inc., 109 FERC ¶ 61,019 (2004).

[166] 2005 Energy Policy Act § 1289; FPA § 203(a)(1)(D).

[167] Order No. 669 at P 84; 18 C.F.R. § 33.1(b)(1).

[168] Order No. 669 at P 85; 18 C.F.R. § 33.1(b)(1).

that facility was operational at a time prior to consummation of the transaction.

The FERC defined the phrase "the time the transaction is consummated" as the point in time when the transaction actually closes and control of the facility changes hands.[169] Further, with regard to the transfer of a generation facility used for interstate wholesale sales over which the FERC has ratemaking jurisdiction, the FERC established a rebuttable presumption that Section 203 applies to any such transfer unless the applicant can demonstrate with substantial evidence that the generator is used exclusively for retail sales.[170] The FERC ruled that a utility acquiring an existing facility that happens to be a QF must obtain Section 203 approval.[171] The FERC explained that, although certain QFs are exempt from Section 203 under regulations implementing PURPA, those PURPA regulations do not exempt a public utility from Section 203 if it acquires a QF.

As for the $10 million threshold that triggers FERC's jurisdiction under Section 203, FERC's general rule will be to use market value. Thus, for transactions between non-affiliates, the FERC will rebuttably presume that the transaction price agreed upon by the parties will be the market value of the subject assets. However, the FERC recognized that, due to the type of facilities that may be transferred and the existence of affiliate transactions, transaction price may not be the best measurement tool in all circumstances. Thus, in evaluating the subject assets to determine whether the $10 million threshold is crossed, the FERC ruled that it will measure value as follows:

	Non-Affiliate	**Affiliate**
Physical Facilities	Transaction price.	Original cost undepreciated as defined in FERC's Uniform System of Accounts.
Contracts	Transaction price.	Total expected nominal contract revenues over the remaining life of the contract.

169 Order No. 669 at P 85; 18 C.F.R. § 33.1(b)(1).

170 Order No. 669 at P 86; 18 C.F.R. § 33.1(b)(1). The FERC explained that, in its experience, utilities do not ordinarily separate the dispatch of their generators for retail sales and wholesale sales. Instead, utilities dispatch their generating facilities to serve all load (retail and wholesale).

171 Order No. 669 at P 87.

	Non-Affiliate	Affiliate
Securities	Transaction price.	If widely traded, market price at time the securities are acquired. If not widely traded, the value of company (based on total undepreciated book value of the company's assets) multiplied by the fraction of the equity securities at issue (determined by dividing the number of equity securities involved in the transaction by the total number of outstanding equity securities for the company).

The FERC also explained that, if a transaction between non-affiliates involves both jurisdictional and non-jurisdictional physical facilities, the "value" of the transaction could be based on a valuation study or the original cost undepreciated of the subject facilities.[172] Likewise, the FERC ruled that if a transaction between non-affiliates involves a wholesale contract and other non-jurisdictional assets, the applicant should rely on the valuation of the contract for determining value.[173] However, the FERC clarified that, with respect to acquisition of securities, whether by affiliates or non-affiliates, the FERC will not allow an applicant to argue that the stock price should be adjusted because it includes value for non-jurisdictional assets.[174]

[iv] Prohibition on Cross-Subsidization to Associate Companies

New Section 203(a)(4) of the FPA requires the FERC to approve a proposed transaction if it finds that the transaction will be consistent with the public interest and not result in cross-subsidization of a non-utility associate company or the pledge or encumbrance of utility assets for the benefit of an associate company unless the FERC determines that the cross-subsidization, pledge, or encumbrance is consistent with

[172] Order No. 669 at PP 116–117. The FERC specifically rejected use of net book value.

[173] Order No. 669 at P 120. The FERC indicated that the market valuation should be consistent with the value that the applicant places on the contract for purposes of its audited financial statements and in keeping with generally accepted accounting principles.

[174] Order No. 669 at P 123.

the public interest. In Order No. 669, the FERC defined "non-utility associate company" to be any associate company in a holding company system other than a public utility or electric utility company that has wholesale or retail customers served under cost-based regulation.[175] This broad definition includes, among other things, a power marketer, a generator that does not have captive customers, a gas marketer, a fuel supply company or company that provides inputs to power production, or a company that is involved in business activities not related to the generation, transmission, distribution, or sale of electricity.

Order No. 669 requires Section 203 applicants to demonstrate how their proposed transactions will not result in cross-subsidization to associate companies.[176] In Order No. 669, the FERC established a new Exhibit M to be included in all Section 203 applications in which an applicant must explain either how the applicant is providing assurances that the proposed transactions will not result in cross-subsidization or improper pledges or encumbrances of utility assets or, if such results would occur, how those results are consistent with the public interest. Alternatively, an applicant may submit a verified statement indicating that "the proposed transaction does not result in, at the time of the transaction or in the future," improper cross-subsidization, pledges, or encumbrances.[177]

The FERC acknowledged that the prohibition on cross-subsidization was similar to the existing concern over the impact of the transaction on rates. Thus, the FERC suggested that Section 203 applicants proffer protection mechanisms to assure ratepayers are protected from the effects of cross-subsidization just as they proffer mechanisms to protect rates.[178] In addition, the FERC advised that applicants should seek to resolve these cross-subsidization matters prior to filing an application under Section 203. The FERC also commented that any protection mechanism would complement, not nullify, mechanisms imposed by a state commission.[179]

---

175 18 C.F.R. § 33.1(b)(2).

176 Order No. 669 at P 164.

177 Order No. 669 at P 169.

178 The FERC indicated the mechanisms may include hold-harmless provisions, rate moratoria, or rate freezes. Order No. 669 P 167. However, each proposal would be addressed on a case-by-case basis.

179 Order No. 669 at P 167.

[v] FERC Procedures for Consideration of Section 203 Applications

The Act requires the FERC to act on certain Section 203 applications on an expedited basis and others within 180 days. In Order No. 669, the FERC did not adopt an exhaustive list of the types of Section 203 applications that would be processed on an expedited basis. However, the FERC indicated that it will generally expedite Section 203 applications "that are not contested, are not mergers, and are consistent with Commission precedent."[180] These include applications proposing the disposition of only transmission facilities, particularly those in which the facilities before and after the transfer are under the control of a FERC-approved RTO or ISO, as well as applications that do not require an Appendix A analysis.[181]

The FERC determined that internal corporate reorganizations that do not present cross-subsidization issues are unlikely to cause anticompetitive effects. Thus, in Order No. 669, the FERC granted blanket authorization for such transactions if they also do not involve a traditional public utility with captive customers.[182] Likewise, the FERC decided to grant blanket authorization for any holding company in a holding company system that includes a transmitting utility or an electric utility company to acquire a foreign utility company. However, if the holding company, its affiliates, it subsidiaries, or associate companies have captive customers in the United States, the blanket authorization is conditioned on the holding company verifying by a duly authorized corporate official of the holding company that the proposed transaction will not result in, "at the time of the transaction or in the future," any improper cross-subsidization, pledge, or encumbrance.[183]

To comply with the expedited requirement, the FERC ruled that applications that involve or require a competitive analysis will be subject to a 60-day notice period and that all other applications will be subject to a 21-day notice period. However, rather than codifying

---

[180] Order No. 669 at P 188.

[181] Order No. 669 at PP 190–191; 18 C.F.R. § 33.11(b).

[182] Order No. 669 at P 192.

[183] Order No. 669 at P 193.

the notice periods, the Commission only adopted them as a policy, in the interest of flexibility.[184]

[10] Relief for Extraordinary Violations

In the Act, Congress clarified that, for any contract entered into prior to June 20, 2001, where the FERC has found that the seller manipulated the energy market resulting in unjust and unreasonable rates, and on that basis has revoked the seller's market-based rate authority, the FERC has exclusive jurisdiction to determine whether it is appropriate to require termination payments for power not delivered.[185] This section applies only to a proceeding pending at the date of enactment of the Act and for which there is not a final, nonappealable order by the Commission or other entity determining the rights of the seller.[186]

[11] Economic Dispatch

The Act creates new Section 222 to the FPA under which Congress requires the FERC to convene joint boards on a regional basis to study security constrained economic dispatch.[187] Under this provision, each state is to nominate a representative for the appropriate regional joint board and the FERC will designate a member of the FERC to serve as chair. Each joint board will consider issues relevant to what constitutes "security constrained economic dispatch" and how such economic dispatch affects or enhances the reliability and affordability of service to customers in the region; the Boards will then make recommendations to the FERC. Within one year of enactment, the FERC is to issue a report and submit it to Congress regarding the recommendations of the joint boards.

§ 1.13 Title XIII—Energy Policy Tax Incentives

[1] Scope

Title XIII of the Act enacts the Energy Tax Incentives Act of 2005 to amend the Internal Revenue Code of 1986 (the "Code") to provide

---

184 Order No. 669 at P 194.

185 2005 Energy Policy Act § 1290(a), (b).

186 2005 Energy Policy Act § 1290(c).

187 2005 Energy Policy Act § 1298. The FERC has opened Docket No. AD05-13-000 for these activities. The FERC held the first meetings of the Joint Boards in November 2005.

a diverse range of tax incentives that benefit all of the major energy industries including electric transmission and generation, natural gas transmission and production, oil refinery, renewable energy, nuclear, and coal. The Act also provides a few modest incentives for individual energy conservation and efficiency from automobiles, home electrical and heating conservation technologies, to home appliances.

[2] Industry Incentives

[a] Electric Transmission and Generation

[i] Fifteen-Year Depreciation for Electrical Transmission Property

The applicable recovery period for assets placed in service under the Modified Accelerated Cost Recovery System is based on the "class life" of the property. The class lives of assets placed in service after 1986 are generally set forth in Revenue Procedure 87-56, 1987-2 C.B. 674 (as clarified and modified by Rev. Proc. 88-22, 1988-1 C.B. 785). Assets used in the transmission and distribution of electricity for sale and related land improvements are given a 20-year recovery period, with a 30-year recovery period for certain assets used in the transmission of electricity for sale and related land improvements.

Section 1308 of the Act shortens the recovery period for transmission property from 20 years to 15 years. These are assets used in the transmission and distribution of electricity for sale and related land improvements. This accelerated depreciation is limited to Section 1245 property used in the transmission at 69 or more kilovolts of electricity for sale, the original use of which commences with the taxpayer after April 11, 2005.[1]

This provision is estimated to cost $1.239 billion over 10 years (which appears to be the largest tax benefit given to the electric transmission sector in the Act).[2]

[ii] Eighty-Four Month Amortization for Pollution Control Facilities

The Act provides accelerated 84-month tax amortization of the cost of air pollution control facilities used in connection with an electric

[1] 2005 Energy Policy Act § 1308 (amending Code § 168).

[2] Congressional Budget Office Cost Estimate, 2005 TNT 144-22, July 25, 2005.

generation plant which is primarily coal fired and which was not in operation before January 1, 1976. For an air pollution control facility used in connection with a plant or other property, the facility must be property that either (1) the construction, reconstruction, or erection of which is completed by the taxpayer after April 11, 2005 (to the extent of the portion of the basis properly attributable to the construction, reconstruction, or erection after that date), or (2) it is acquired by the taxpayer after April 11, 2005, and the original use starts after that date. Certain certified air pollution control facilities placed in service before January 1, 1976, and used in connection with certain primarily coal fired electric plants generally continue to be eligible for amortization over 60 months.[3]

This provision is estimated to cost $1.147 billion.[4]

[iii] Five Year Carryback Net Operating Losses for Electric Transmission Equipment

A net operating loss (NOL) is the amount by which a taxpayer's allowable deductions exceed the taxpayer's gross income. A carryback of an NOL generally results in the refund of federal income tax for the carryback year, just as a carryover of an NOL reduces federal income tax for the carryover year. Generally, Internal Revenue Code Section 172 provides that an NOL may be carried back two years and carried over 20 years to offset taxable income in such years. The ordering rules generally require NOLs to be applied to the earliest of the taxable years to which the loss may be carried.

Under the Act, in recognition of the losses incurred by some electrical utilities in recent years, tax losses of electric utilities in 2003, 2004, and 2005 may be carried back five years (rather than the usual two) to the extent of the utility's investment in electric transmission property or pollution control equipment, thus generating tax refunds. The election may be made during any taxable year ending after December 31, 2005, and before January 1, 2009. An electing taxpayer must specify to which loss year the election applies. The portion of the loss year NOL which the election was made is limited to 20% of the amount of the investment outlay in either the electric transmission property or pollution control facilities in the year prior to the year

[3] 2005 Energy Policy Act § 1309 (amending Code § 169).

[4] Congressional Budget Office Cost Estimate, 2005 TNT 144-22, July 25, 2005.

in which the NOL election occurred. An investment in electric transmission property qualifies if it is a capital expenditure made by the taxpayer which is attributable to electric transmission property used by the taxpayer in the transmission at 69 or more kilovolts of electricity for sale.

The conference report to the Act indicates that there is no requirement that the transmission property or pollution control facilities be placed in service in the year in which the capital expenditures are incurred.[5] However, it is intended that qualifying investment under the provision includes only capital expenditures to which the taxpayer is committed and with respect to property which the taxpayer intends to ultimately place in service in the taxpayer's trade or business.[6] This provision is estimated to cost $52 million.[7]

[iv] Deferred Gain From Sale of Transmission Assets

A taxpayer selling property recognizes gain to the extent the sales price exceeds the basis in the property. The recognized gain is subject to current income tax unless the gain is deferred or not recognized under another provision in the Internal Revenue Code. Code Section 451(i) allows an election to recognize gain from qualifying electric transmission transactions ratably over an eight-year period beginning in the year of sale if the amount realized from such sale is used to purchase exempt utility property within the applicable period (which is four years after the close of the taxable year in which the qualifying electric transmission transaction occurs). A qualifying electric transmission transaction is the sale or other disposition of property used by the taxpayer in the trade or business of providing electric transmission services, or an ownership interest in such an entity, to an independent transmission company prior to January 1, 2007. If the amount realized exceeds the amount used to purchase reinvestment property, any realized gain is recognized to the extent of such excess in the year of the qualifying electric transmission transaction.

The Act extends present law for electric utilities that sell their transmission assets to a FERC-approved independent company by

---

5 Section D.6 of the Technical Explanation of the Conference Agreement for H.R. 6, Title XIII, "Energy Tax Incentives Act of 2005."

6 2005 Energy Policy Act § 1311 (amending Code § 172).

7 Congressional Budget Office Cost Estimate, 2005 TNT 144-22, July 25, 2005.

allowing payment of tax on the gains over eight years for transactions prior to January 1, 2008.[8] This provision is estimated to cost $19 million.[9]

[b] Electric Cooperative Income Relief

Section 501(c)(12) of the Internal Revenue Code provides that tax-exempt rural electric cooperatives are required to have 85% of the cooperative's income consist of amounts collected from members of the cooperative to meet losses and expenses of providing service to its members (the "85/15 test"). This test had made it difficult for cooperatives to participate in electricity market deregulation and open access transmission of electricity. The 85/15 test was previously modified to exclude income received or accrued by tax-exempt rural electric cooperatives from open access electric energy transmission or distribution services, any nuclear decommissioning transaction, and any asset exchange or conversion transaction under Code Section 501(c)(12). The Act makes permanent this exclusion, which otherwise would have expired at the end of 2006. The Act also eliminates the sunset date for the rule that permits taxable electric cooperatives to treat the receipt or accrual of income from load loss transactions as income from patrons who ware members of the cooperative.[10] This provision is estimated to cost $277 million.[11]

[c] Natural Gas Transmission and Production

[i] Accelerated Depreciation for New Investment in Natural Gas Distribution Lines

The Act provides that new investments in natural gas distribution lines will be entitled to accelerated depreciation deductions, being written off over 15 years instead of 20. To qualify, the new natural gas distribution lines must be placed in service after April 11, 2005, and before January 1, 2011. The provision does *not* apply to property subject to a binding contract on or before April 11, 2005.[12] This provision is estimated to cost $1.019 billion.[13]

---

[8] 2005 Energy Policy Act § 1305 (amending Code § 451).

[9] Congressional Budget Office Cost Estimate, 2005 TNT 144-22, July 25, 2005.

[10] 2005 Energy Policy Act § 1304 (amending Code § 501(c)(12)).

[11] Congressional Budget Office Cost Estimate, 2005 TNT 144-22, July 25, 2005.

[12] 2005 Energy Policy Act § 1325 (amending Code § 168).

[13] Congressional Budget Office Cost Estimate, 2005 TNT 144-22, July 25, 2005.

[ii] Seven-Year Depreciation for Natural Gas Gathering Lines

In three recent cases, appellate courts have held that natural gas gathering lines owned by nonproducers are depreciable over seven years (rather than 15 years as the Internal Revenue Service (IRS) had argued).[14] The Act confirms this conclusion. The original use of the property must begin with the taxpayer.[15] This provision is estimated to cost $16 million.[16]

[iii] Arbitrage Rules Do Not Apply for Prepayment of Natural Gas

Under Code Section 148, arbitrage is the profit that results from investing the proceeds of tax-exempt bonds in higher yielding taxable securities; such arbitrage generally must be rebated to the United States government. The Act creates a safe harbor from the arbitrage rules to allow public utilities to finance prepayments for natural gas with tax exempt bond proceeds if the natural gas is used to supply the utility's customers. Accordingly, this provision should permit utilities to purchase natural gas supplies for their customers at lower prices.[17] This provision is estimated to cost $53 million.[18]

[iv] Amortization of Geological and Geophysical Expenditures

The Act allows two-year amortization of geological and geophysical costs incurred in connection with oil and gas exploration in the United States. For abandoned property, any remaining basis may no longer be recovered in the year of abandonment.[19] This provision is estimated to cost $974 million.[20]

---

[14] *See* Clajon Gas Co., L.P. v. Commissioner, 354 F.3d 786 (8th Cir. 2004), *rev'g* 119 T.C. 197 (2002); Saginaw Bay Pipeline Co. v. United States, 338 F.3d 600 (6th Cir. 2003), *rev'g* 88 A.F.T.R.2d 2001-6019 (E.D. Mich. 2001); Duke Energy v. Commissioner, 172 F.3d 1255 (10th Cir. 1999), *rev'g* 109 T.C. 416 (1997).

[15] 2005 Energy Policy Act § 1328 (amending Code § 168).

[16] Congressional Budget Office Cost Estimate, 2005 TNT 144-22, July 25, 2005.

[17] 2005 Energy Policy Act § 1327 (amending Code § 148).

[18] Congressional Budget Office Cost Estimate, 2005 TNT 144-22, July 25, 2005.

[19] 2005 Energy Policy Act § 1329 (amending Code § 167).

[20] Congressional Budget Office Cost Estimate, 2005 TNT 144-22, July 25, 2005.

[d] Oil Refining

[i] Expanded Smaller Refiner Exception to Oil Depletion Deduction

Previously, a producer could qualify as an independent producer (and thus obtain percentage depletion) if its refining operations did not exceed 50,000 barrels on any day in the taxable year. The Act increases the 50,000 barrel per day cap to 75,000. Further taxpayer relief is provided by basing the 75,000 limitation on an average daily production for the taxable year.[21] This provision is estimated to cost $158 million.[22]

[ii] Incentive for Liquid Fuels Refiners

Refiners of liquid fuels are given an incentive to make investments in qualified refinery property by being able to expense half their purchase. The remaining costs will continue to be depreciated as 10-year property under MACRS. For existing refineries to receive the benefit, the equipment must increase capacity by at least 5% or increase the throughput of certain qualified fuels by at least 25% on an average daily basis. Qualified fuels include oil from shale and tar sands. Refineries must report to the IRS on production and output to become eligible for the expensing. The allowance sunsets after December 31, 2011.[23] This provision is estimated to cost $406 million.[24]

[iii] Cooperative Expensing of EPA Sulfur Compliance Costs

Small refiner cooperatives will now be permitted to pass through to their owners a deduction for capital costs incurred to comply with EPA sulfur regulations.[25] This provision is estimated to cost $7 million.[26]

---

[21] 2005 Energy Policy Act § 1328 (amending Code § 613A).

[22] Congressional Budget Office Cost Estimate, 2005 TNT 144-22, July 25, 2005.

[23] 2005 Energy Policy Act § 1323 (amending Code § 179C).

[24] Congressional Budget Office Cost Estimate, 2005 TNT 144-22, July 25, 2005.

[25] 2005 Energy Policy Act § 1324 (amending Code § 179B).

[26] Congressional Budget Office Cost Estimate, 2005 TNT 144-22, July 25, 2005.

[e] Renewable Energy Industry

[i] Extension of Placed in Service Date and Expansion of Sources

An income tax credit is allowed for the production of electricity from qualified facilities sold by the taxpayer to an unrelated person under Section 45 of the Internal Revenue Code. The base amount of the credit is 1.5 cents per kilowatt-hour (indexed for inflation) of electricity produced. The amount of the credit is 1.9 cents per kilowatt-hour for 2005. A taxpayer may claim credit for the 10-year period commencing with the date the qualified facility is placed in service. The credit is reduced for grants, tax-exempt bonds, subsidized energy financing, and other credits. The amount of the credit is subject to phase out as the market price of electricity (or refined coal for the refined coal production credit) exceeds certain thresholds.

The "placed in service" deadline for facilities to qualify for these tax credits under Code Section 45 was extended by two years, to December 31, 2007, under the Act. The new deadline applies to wind, closed-loop biomass, open-loop biomass, geothermal, small irrigation power, landfill gas, and trash combustion facilities. The deadline was not extended for solar facilities (terminating December 31, 2005) or refined coal facilities (December 31, 2008). This two year extension was a disappointment, since project development cycles can take two years or more for some renewables projects; the industry had hoped for a three year extension or more. However, this is the first time Congress has acted to extend the Section 45 tax credit before it expired, thus hopefully avoiding a repeat of the "boom and bust" cycle that has characterized renewable energy development.

The list of sources qualifying for such credits was expanded again to include incremental hydropower, new units at existing trash combustion facilities, coke and coke gas facilities, and coal owned by Indian tribes. Also, previously the credit had applied to electricity produced over the first five or 10 years of production, depending on the type of energy resource; now the credit applies for the first 10 years in almost all cases (but seven for Indian coal and only four for coke and coke gas facilities).[27] These various changes to Section 45

27 2005 Energy Policy Act § 1301 (amending Code § 45).

of the Internal Revenue Code are estimated to cost the Treasury $2.747 billion over 10 years.[28]

[ii] Tax Credit Bonds Supporting Renewable Energy

Interest on state and local government bonds generally is excluded from gross income for federal income tax purposes if the proceeds of the bonds are used to finance direct activities of these governmental units or if the bonds are repaid with revenues of the governmental units. In an interesting twist on distributing tax-exempt interest, the Act authorizes the issuance of $800 million of tax-credit bonds brought to the market after December 31, 2005, and before December 31, 2007, that supports renewable investment.

These Clean Renewable Energy Bonds (CREBs) are defined as bonds issued by qualified issuers if, in addition to other requirements, 95% of the proceeds are used to finance facilities qualifying for tax credit under Section 45 of the Internal Revenue Code. Qualified issuers include governmental bodies and mutual or cooperative electric companies. CREBs are not interest-bearing obligations. Instead, the holder of a CREB, on a credit allowance date is entitled to a tax credit. The amount of the credit is determined by multiplying the bond's credit rate by the face amount on the holder's bond. The credit rate on the bond is a rate that permits issuance of CREBs without discount and interest cost to the qualified issuer. The credit accrues quarterly and is includible in gross income, similar to an interest payment. The credit can be claimed against regular income tax liability and alternative minimum tax liability.[29] This provision is estimated to cost $411 million.[30]

[iii] Extension of Income Tax Credit, Excise Tax Credit, and Payments for Biodiesel and New Renewable Diesel Credit

Section 40A of the Internal Revenue Code provides an income tax credit for biodiesel and qualified biodiesel mixtures, the biodiesel fuels credit. The biodiesel fuels credit is equal to the biodiesel mixture credit plus the biodiesel credit. Any credits taken are includible in gross

---

28 Congressional Budget Office Cost Estimate, 2005 TNT 144-22, July 25, 2005.

29 2005 Energy Policy Act § 1303 (amending Code § 54).

30 Congressional Budget Office Cost Estimate, 2005 TNT 144-22, July 25, 2005.

income. The provisions for the biodiesel fuels credit and the biodiesel excise tax credit are coordinated together.

The Act extends the biodiesel and qualified biodiesel income tax credits, excise tax credit, and payment provisions through December 31, 2008, for biodiesel and qualified biodiesel mixtures, and biodiesel fuels credit. The biodiesel mixture credit is 50 cents for each gallon of biodiesel used by the taxpayer in the production of a qualified biodiesel mixture. The biodiesel credit is 50 cents for each gallon of biodiesel which is not in a mixture with diesel fuel (100% biodiesel or B-100), subject to certain requirements.

Code Section 6426(c) provides an excise tax credit for biodiesel mixtures. The credit is 50 cents for each gallon of biodiesel used by the taxpayer in producing a biodesel mixture for sale or use in a trade or business of the taxpayer. If any person produces a biodiesel fuel mixture in that person's trade or business, a payment is made to that person equal to the biodiesel mixture credit under Code Section 6426(e). To the extent the biodiesel fuel mixture credit exceeds the Code Section 4081 liability of a person, such person is paid an amount equal to the biodiesel fuel mixture credit.

The Act adds a similar credit for renewable diesel (diesel fuel derived from biomass using a thermal depolymerization process). Renewable diesel must meet the requirements of ASTM D975 or D396 and meet certain requirements established by the EPA under the Clean Air Act. There is no special credit for small producers of renewable diesel.[31] This provision is estimated to cost $194 million.[32]

[iv] Smaller Producer Biodiesel and Ethanol Credit

The Act adds a credit for small agri-biodiesel producers under the biodiesel fuels credit. The credit is 10-cents-per-gallon for up to 15 million gallons of agri-biodiesel produced by small producers. The credit sunsets after December 31, 2008.[33] The Act also increases the limit of production capacity for small ethanol producers from 30

[31] 2005 Energy Policy Act §§ 1344, 1346 (amending Code §§ 40A, 6426, 6427).

[32] Congressional Budget Office Cost Estimate, 2005 TNT 144-22, July 25, 2005.

[33] 2005 Energy Policy Act § 1345 (amending Code § 40A).

million gallons to 60 million gallons per year.[34] This provision is estimated to cost $181 million.[35]

[f] Nuclear

[i] Expansion of Qualified Nuclear Decommission Fund Deductions

A qualified nuclear decommissioning fund (a "qualified fund") is a segregated fund established by a taxpayer that is used exclusively for the payment of decommissioning costs, taxes on fund income, management costs of the fund, and for making investments. Unlike normal additions to reserves, amounts contributed to a qualified nuclear decomissioning fund are tax deductible, subject to various limitations. Funds withdrawn to pay for decommissioning costs are included in the taxpayer's income.

The Act repeals the prior law limitation that contributions to a qualified fund are deductible in the year made only to the extent that these amounts were collected as part of the cost of service to ratepayers. Thus, the deduction for contributions to a qualified nuclear decommission fund under Code Section 468A is expanded to unregulated merchant producers.

The Act also repeals the prior law limitation that deductions are available only for post-1984 decommissioning costs. Therefore, any taxpayer is permitted to accumulate an amount sufficient to cover the present value of 100% of a nuclear powerplant's estimated decommissioning cost in a qualified fund. The Act even allows the deduction of contributions in excess of limitations specified in private letter rulings governing the taxpayer's decommissioning fund, up to the present value of total nuclear decommissioning costs with respect to a nuclear power plant previously excluded under Code Section 468A(d)(2)(A).[36] These changes are estimated to cost the Treasury $1.293 billion over 10 years.[37]

---

[34] 2005 Energy Policy Act § 1347 (amending Code § 40).

[35] Congressional Budget Office Cost Estimate, 2005 TNT 144-22, July 25, 2005.

[36] 2005 Energy Policy Act § 1310 (amending Code § 468A(d)(2)(A).

[37] Congressional Budget Office Cost Estimate, 2005 TNT 144-22, July 25, 2005.

[ii] Production Tax Credit for Advanced Nuclear Facilities

The Act adds a production tax credit under Code Section 45J for a qualifying advanced nuclear facility at a rate generally of 1.8 cents per kilowatt-hour for electricity produced over an 8-year period when the facility is placed in service. The amount a taxpayer may claim in any tax year is capped at $125 million per 1,000 megawatts of allocated capacity in any one year. The production tax credit amount is also subject to a further limitation equal to the ratio of allocated capacity (determined by the Secretary of Energy) over the rated nameplate capacity of the facility.

An advanced nuclear facility is any facility for the production of electricity, the reactor design for which was approved after 1993 by the Nuclear Regulatory Commission. For this purpose, a qualifying advanced nuclear facility does not include any facility for which a substantially similar design for a facility of comparable capacity was approved before 1994.[38] This provision is estimated to cost the Treasury $278 million.[39]

[g] Coal

[i] Investment Tax Credits for Clean Coal

The Act creates three different tax credits related to coal operations.

First, the Act establishes a 20% credit for qualified gasification projects. Qualified gasification projects convert coal, petroleum residue, biomass, or other materials recovered for their energy or feedstock value into a synthesis gas composed primarily of carbon monoxide and hydrogen for direct use or subsequent chemical or physical conversion. The Secretary may certify a maximum of $1 billion in qualified investment as eligible for credit with respect to any single project.

Second, the Act creates a 20% credit for integrated gasification combined cycle (IGCC) projects. The credits from IGCC projects require certification by the Secretary of Treasury in consultation with the Secretary of Energy. The definition of credit eligible investments

---

38 2005 Energy Policy Act § 1306.

39 Congressional Budget Office Cost Estimate, 2005 TNT 144-22, July 25, 2005.

includes only investments in property associated with the gasification of coal, including any coal handling and gas separation equipment. Therefore, investments in equipment that could operate by drawing fuel directly from a natural gas pipeline do not qualify for the credit.

Lastly, a 15% credit is given for other advanced coal-based projects that produce electricity.[40]

These credits are only available, however, if the project is certified by the Secretary of Treasury in consultation with the Department of Energy, and there are caps on the amounts that can be certified under each category (a combined $1.65 billion). Although the details of the certification process are not yet clear, interested parties may need to apply for certification promptly to obtain any credits. The Secretary may allocate $800 million of credits to IGCC projects and $500 million of credits to projects using other advanced coal-based technologies and up to $350 million for industrial gasification. Certifications are issued using a competitive bidding process.[41] This provision is estimated to cost $1.612 billion.[42]

[h] Credit for Facilities Producing Coke or Coke Gas

The Act adds a production credit for qualified facilities that produce coke or coke gas. According to the conference report to the Act, a single facility for the production of coke or coke gas is generally composed of multiple coke ovens or similar structures.[43] The $3 credit is indexed for inflation and may not exceed an average barrel-of-oil equivalent of 4,000 barrels per day (i.e., each facility is limited to $4.3 million per year). Qualified facilities must have been placed in service before January 1, 1993, or after June 30, 1998, and before January 1, 2010. The production credit may be claimed with respect to coke and coke gas produced and sold during period beginning on the later of January 1, 2006, or the date such facility is placed in service and ending on the date which is four years after such period began. Therefore, the credit expires in January 1, 2010, or four years after

[40] 2005 Energy Policy Act § 1307 (amending Code §§ 48A, 48B).

[41] 2005 Energy Policy Act § 1307 (amending Code §§ 48A, 48B).

[42] Congressional Budget Office Cost Estimate, 2005 TNT 144-22, July 25, 2005.

[43] Section 5 of the Description and Technical Explanation of the Conference Agreement to H.R. 6, Title XIII, "Energy Tax Incentive Act of 2005," July 27, 2005.

the facility was placed in service, whichever is later.[44] This provision is estimated to cost $101 million.[45]

[i] Credit for Producing Fuel From a Nonconventional Source

Under the Act, the IRC Section 29 credit for producing fuel from non-conventional sources is moved so that it becomes a general business credit, thereby allowing unused credits to be carried back one year and forward 20 years.[46] This provision is estimated to cost $88 million.[47]

[j] Miscellaneous Incentives for Businesses

[i] Installation of Qualified Fuel Cells, Microturbine Power Plants, and Solar

The Act provides a 30% business energy credit for purchased qualified fuel cell power plants for businesses and a 10% credit for qualified stationary microturbine power plants. A qualified fuel cell power plant generally is an integrated system composed of a fuel cell stack assembly and associated balance of plant components that (1) convert fuels into electricity using electrochemical methods, (2) has electrical generation efficiency of greater than 30%, and (3) generates at least 0.5 kilowatts of electricity. A qualifying stationary microturbine power plant is an integrated system that includes a gas turbine engine, a combustor, a recuperator or regenerator, a generator or alternator, and associated balance of plant components that converts fuel into electricity and thermal energy.[48]

The Act increases the 10% credit to 30% in the case of solar energy property. Additionally, the Act provides that equipment that uses fiber-optic distributed sunlight to illuminate the inside of a structure is solar energy property.

The credits apply to periods after December 31, 2005, and before January 1, 2008, for property placed in service in taxable years ending

---

[44] 2005 Energy Policy Act § 1321 (adding subsection (h) to Code § 29).

[45] Congressional Budget Office Cost Estimate, 2005 TNT 144-22, July 25, 2005.

[46] 2005 Energy Policy Act § 1322 (moving Code § 29 to § 45K).

[47] Congressional Budget Office Cost Estimate, 2005 TNT 144-22, July 25, 2005.

[48] 2005 Energy Policy Act § 1336 (amending Code § 48).

after December 31, 2005, under rules similar to rules of Section 48(m) of the Internal Revenue Code (as in effect on the day before the date of enactment of the Revenue Reconciliation Act of 1990).[49] This provision is estimated to cost $222 million.[50]

[ii] Deduction for Energy Efficient Commercial Buildings

Energy efficient commercial buildings that reduce annual energy and power consumption by 50% compared to the American Society of Heating, Refrigerating, and Air Conditioning engineers standard are allowed to deduct the cost of the energy efficient property installed with a maximum deduction of $1.80 per square foot of the building. In addition, a partial deduction limited to 60 cents per square foot would be provided for building subsystems. Certain certification requirements must be met in order to qualify for the deduction. The provision is effective for property placed in service after December 31, 2005, and prior to January 1, 2008.[51] This provision is estimated to cost $243 million.[52]

[iii] Installation of Alternative Fuel Refueling Property

A 30% credit is given under the Act for the cost of installing clean fuel vehicle refueling property to be used in a trade or business of the taxpayer or installed at the principal residence of the taxpayer subject to dollar limitations (retail ($30,000) and residential ($1,000)). Clean fuel refueling property is property used for the storage or dispensing of a clean burning fuel, if the storage or dispensing is at the point at which the fuel is delivered into the fuel tank of a motor vehicle. Clean fuels are any fuel at least 85% of the volume of which consists of ethanol, natural gas, compressed natural gas, liquefied natural gas, liquefied petroleum gas, and hydrogen and any mixture of diesel fuel and biodiesel containing at least 20% biodiesel.

The part of the credit that relates to property that is of a type subject to depreciation is treated as a portion of the general business credit. The basis in the property is reduced by the amount of the credit and

---

[49] 2005 Energy Policy Act § 1337 (amending Code § 48).

[50] Congressional Budget Office Cost Estimate, 2005 TNT 144-22, July 25, 2005.

[51] 2005 Energy Policy Act § 1331 (adding new Code § 179D).

[52] 2005 Energy Policy Act § 1342 (amending Code § 30C).

the taxpayer may not claim deductions under Code Section 179A with respect to property for which the credit is claimed. The remainder of the credit is allowable to the extent of the excess of the regular tax (reduced by certain other credits) over the alternative minimum tax for the taxable year. The provisions of Code Section 30C are generally applicable with respect to property placed in service after December 31, 2005, and before January 1, 2008.[53] The provision is estimated to cost $71 million.[54]

[3] Consumer Conservation and Efficiency Incentives

[a] Alternative Technology Vehicle Credits

The Act provides a credit for new qualified fuel cell vehicles, qualified alternative fuel motor vehicles, new qualified hybrid motor vehicles, and advanced lean burn technology motor vehicles. The credit is generally allowed to the owner (which includes the lessor). A vehicle must be used predominately in the United States to qualify for the credit.

A qualifying fuel cell vehicle is a motor vehicle that is propelled by power derived from one or more cells which convert chemical energy directly into electricity by combining oxygen with hydrogen fuel (which is stored on board the vehicle and may or may not require reformation prior to use). The amount of credit for the purchase of a fuel cell vehicle is determined by a base credit amount that depends upon the weight class of the vehicle. For automobiles or light trucks, an additional credit amount is added to the credit that is based upon the rated fuel economy of the vehicle compared to a base fuel economy.

The credit for the purchase of a new alternative fuel vehicle would be 50% of the incremental cost of such vehicle, plus an additional 30% if the vehicle meets certain emissions standards with the credit being limited depending upon the weight of the vehicle. An alternative fuel is compressed natural gas, liquefied natural gas, liquefied petroleum gas, hydrogen, and any liquid fuel that is at least 85% methanol. Qualifying alternative fuel motor vehicles are vehicles that generally

53 2005 Energy Policy Act § 1342 (amending Code § 30C).

54 Congressional Budget Office Cost Estimate, 2005 TNT 144-22, July 25, 2005.

operate only on qualifying alternative fuels and are incapable of operating on gasoline or diesel.

A qualifying hybrid vehicle is a motor vehicle that draws propulsion energy from on-board sources of stored energy which include both an internal combustion engine or heat engine using combustible fuel and a rechargeable energy storage system such as batteries.

A qualifying advanced lean burn technology motor vehicle incorporates direct injection, achieves at least 125% of the 2002 model year city fuel economy, and for 2004 and later model vehicles meets or exceeds certain EPA emissions standards.

The amount of the credit for a qualifying hybrid vehicle and an advanced lean burn technology motor vehicle equals a fuel economy credit amount that varies with the rated fuel economy of the vehicle (compared to a 2002 model year baseline) plus a conservation credit based on the estimated lifetime fuel savings of a qualifying vehicle (compared to a comparable 2002 model year vehicle).

Certain production limitations are placed on each manufacturer of a qualified hybrid motor vehicle and an advanced lean burn technology motor vehicle.

These provisions apply to vehicles placed in service after December 31, 2005. A qualified fuel cell motor vehicle must be placed in service, before January 1, 2015; qualified hybrid motor vehicles that are automobiles and light trucks and advanced lean-burn technology vehicles must be placed in service before January 1, 2011; qualified hybrid motor vehicles that are medium and heavy trucks must be placed in service before January 1, 2010; and qualified alternative fuel motor vehicles must be placed in service before January 1, 2011.[55] This provision is estimated to cost $874 million.[56]

[b] Incentives for Energy Efficient Homes

[i] Existing Homes

The Act provides a 10% investment tax credit for certain qualified energy efficiency improvements made to existing homes. To claim the credit, generally the improvements must meet or exceed certain

55 2005 Energy Policy Act §§ 1341, 1348 (adding new Code § 30B).

56 Congressional Budget Office Cost Estimate, 2005 TNT 144-22, July 25, 2005.

criteria for such component established by the 2000 International Energy Conservation Code and pass certain other requirements. Also, specific credits are given for purchases of advanced main air circulating fans, natural gas, propane, or oil furnaces or hot water boilers, and other qualified energy efficient property. The credit allowed may not exceed $500 in total across all taxable years, and no more than $200 of such credit may be attributable to expenditures on windows. The property must be placed in service after December 31, 2005, and before January 1, 2008.[57] This provision is estimated to cost $556 million.[58]

[ii] New Homes

Contractors can receive a credit for construction of qualified new energy efficient homes that meet certain energy efficiency standards and other requirements. The credit applies to homes purchased after December 31, 2005 and prior to January 1, 2008, and that are substantially completed after December 31, 2005.[59] This provision is estimated to cost $28 million.[60]

[c] Home Modifications

The Act provides a personal tax credit, equal to 30% of qualifying expenditures, for the purchase of qualified photovoltaic property or qualified solar water heating property used exclusively for a residence within the United States and used other than for heating swimming pools and hot tubs. The credit is equal to 30% of qualifying expenditures, with a maximum credit for each of these systems of property of $2,000. A 30% credit is also provided for the purchase of a qualified fuel cell power plant which is an integrated system comprised of a fuel cell stack assembly and associated balance of plant components generating electricity meeting certain efficiency and kilowatts standards for a residence in the United States. The provision applies to property placed in service after December 31, 2005 and prior to January 1, 2008.[61] This provision is estimated to cost $31 million.[62]

---

57 2005 Energy Policy Act § 1333 (adding new Code § 25C).

58 Congressional Budget Office Cost Estimate, 2005 TNT 144-22, July 25, 2005.

59 2005 Energy Policy Act § 1332 (adding new Code § 45L).

60 Congressional Budget Office Cost Estimate, 2005 TNT 144-22, July 25, 2005.

61 2005 Energy Policy Act § 1335 (adding new Code § 25D).

62 Congressional Budget Office Cost Estimate, 2005 TNT 144-22, July 25, 2005.

[d] Energy Efficient Home Appliances

Manufacturers receive a tax credit for the production of certain energy-efficient dishwashers, clothes washers and refrigerators. The amount of the credit varies by appliance, has differing standards from appliance to appliance, and generally is calculated based upon efficiency benchmarks. The provision is effective for appliances manufactured after December 31, 2005 and before January 1, 2008.[63] This provision is estimated to cost $180 million.[64]

§ 1.14 Title XIV—Miscellaneous

[1] Scope

The Energy Policy Act of 2005 includes miscellaneous provisions designed to advance Congress's goals of fostering renewable energy technology and achieving energy self-sufficiency.

[2] Energy Production Incentives

Under Section 1402 of the Act, a state may provide a tax credit or other tax-related incentive for the production within the state of electricity from coal mined within the state and used in a facility in the state, subject to certain limitations.[1] The Act further provides that the provision of such credits is to be considered a reasonable regulation of commerce, thus insulating the state's action from legal challenges under the dormant commerce clause.[2]

[3] Soybean Oil for Use in Electric Transformers

Section 1403 of the Act[3] provides regulatory relief for use of soybean oil as thermal insulation in electric transformers, by exempting such oil from regulation by the Environmental Protection Agency under the Edible Oil Regulatory Reform Act.[4]

---

[63] 2005 Energy Policy Act § 1341 (adding new Code § 45M).

[64] Congressional Budget Office Cost Estimate, 2005 TNT 144-22, July 25, 2005.

[1] 2005 Energy Policy Act § 1402(a)–(b).

[2] 2005 Energy Policy Act § 1402(c).

[3] 2005 Energy Policy Act § 1403.

[4] 33 U.S.C. § 2720(a)(1)(B).

[4] National Priority Project Designation

The Act establishes a National Priority Project Designation to reward organizations that have advanced the field of renewable energy technology and contributed to North American energy independence.[5] Designations will fall into the following four categories: (1) wind and biomass energy generation projects, (2) photovoltaic and fuel cell energy generation projects, (3) energy efficient building and renewable energy projects, and (4) first-in-class renewable energy projects. An organization receiving the designation may publicize the designation and identify itself as a National Priority Project.

[5] Studies Directed

Title XIV of the Act directs the Secretary of Energy to conduct studies of the health impact on persons living near petrochemical and oil refinery facilities.[6] The Act also directs the Secretary to study the application of radiation to petroleum to refine petroleum products, with the goal of increasing the economic yield from each barrel of oil.[7]

[6] Program on Oxygen-Fuel Systems

Section 1407 of the Act establishes, under the auspices of the Department of Energy (DOE), a program on oxygen-fuel systems[8] that must include the renovation of at least one existing large unit (100 megawatts or more) and one existing small unit (10 to 50 megawatts), as well as the construction of one new large unit and one new small unit.[9]

[7] North American Energy Freedom

Section 1421 of the Act creates the "Set America Free Act of 2005" or "SAFE Act."[10] Under the SAFE Act, Congress establishes the

[5] 2005 Energy Policy Act § 1405.

[6] 2005 Energy Policy Act § 1404.

[7] 2005 Energy Policy Act § 1406.

[8] An "oxygen-fuel system" is defined as a system that benefits from the efficient use of oil, gas, coal and biomass combustion using substantially pure oxygen, with high flame temperatures and the exclusion of air from the boiler, in industrial or electric utility steam-generating units. 2005 Energy Policy Act § 1407(c)(2).

[9] 2005 Energy Policy Act § 1407.

[10] 2005 Energy Policy Act §§ 1421–1424.

United States Commission on North American Energy Freedom to make recommendations to the President and Congress for a coordinated and comprehensive North American energy policy that will achieve energy self-sufficiency by 2025 within Canada, Mexico, and the United States. The Commission will be temporary in nature, terminating 90 days after the submission of its final report, which is to be submitted by August 8, 2006. It will have 16 members appointed by the President, subject to certain limitations.

§ 1.15 Title XV—Ethanol and Motor Fuels

[1] Scope

The Act contains new provisions intended to promote the use of ethanol and other renewable fuels in motor vehicle fuels sold or dispensed in the contiguous United States.

[2] Renewable Fuel Program

Section 1501(a) of the Act amends Section 211 of the Clean Air Act to require the inclusion of specific aggregate volumes of renewable fuel in motor vehicle fuel sold or dispensed in the contiguous United States. The Act directs the Administrator of the Environmental Protection Agency (EPA), within one year of enactment, to promulgate regulations to ensure enforcement and compliance. The applicable volumes of renewable fuel in motor vehicle fuel for years 2006-2012 are to be determined in accordance with a Schedule set forth in the Act, which sets the applicable volume at 4.0 billion gallons in 2006, rising to 7.5 billion gallons by 2012. Thereafter, the Administrator of the EPA, in coordination with the Secretaries of Agriculture and Energy, will determine the applicable volume, which must contain a minimum of 250 million gallons derived from cellulosic biomass.[1]

The Act expresses a preference for cellulosic biomass ethanol and waste derived ethanol by specifying that one gallon of such fuels will count as 2.5 gallons for compliance purposes.[2]

Section 1501(a) of the Act also establishes a renewable fuel credit program, under which any person that refines, blends, or imports gasoline containing a quantity of renewable fuel greater than the

---

[1] 2005 Energy Policy Act § 1501(a).

[2] 2005 Energy Policy Act § 1501(a).

applicable volume in a given year receives "credits" valid for 12 months. An entity receiving such credits may use the credits to reduce its applicable volume requirement for that year or transfer all or a portion of the credits to another for purposes of compliance. Any entity unable to satisfy the applicable volume requirement (by generating or purchasing sufficient credits) may carry forward a renewable fuel deficit to the following calendar year, provided it achieves compliance in that year.[3]

Small refineries are temporarily exempt from the applicable volume requirements.[4] The exemption is set to expire in 2011, with the possibility of an extension by the Secretary of Energy or the Administrator of the EPA. A small refinery may choose to waive the exemption, in which case it will become subject to the applicable volume requirement and eligible to generate credits for renewable fuel generated in excess of the annual requirement.[5]

Waivers of the applicable volume requirement are available under the Act, on petition by one or more states. Upon receipt of such a petition, the Administrator, in consultation with the Secretaries of Agriculture and Energy, may waive the requirement in whole or in part by reducing the national quantity of renewable fuel required, if the Administrator determines that (1) implementation of the requirement would severely harm the economy or environment of a state, region or the United States, or (2) there is inadequate domestic supply. Once granted, the waiver is valid for one year, with a possibility of renewal. The Act also makes special provision for a potential waiver in 2006 if the Administrator determines that implementation of the applicable volume requirement will have a significant adverse effect on consumers in 2006 on a national, regional, or statewide basis.[6]

[3] Methyl Tertiary Butyl Ether (MTBE)

One of the major impediments to passage of a comprehensive energy bill prior to 2005 was whether Congress should indemnify fuel

---

3 2005 Energy Policy Act § 1501(a).

4 A "small refinery" is defined as defined as a refinery in which the average aggregate daily crude oil throughput for a calendar year is 75,000 barrels or less. 2005 Energy Policy Act § 1501(a).

5 2005 Energy Policy Act § 1501(a).

6 2005 Energy Policy Act § 1501(a).

companies for having used MTBE as a fuel additive to address requirements under the Clean Air Act. The Act does not indemnify fuel companies for having used MTBE. However, in Section 1502 of the Act, Congress acknowledges that MTBE has been used since 1979 as an octane booster or anti-knocking agent and that the fuel industry responded to the fuel oxygen standard under the Clean Air Act Amendments of 1990 by using MTBE.[7] The Act also provides that claims and legal actions associated with the actual or threatened contamination of MTBE filed after August 8, 2005, may be removed to a federal district court.[8]

[4] Elimination of Oxygen Content Requirement for Reformulated Gasoline

Section 1504 of the Act amends Section 212(k) of the Clean Air Act by eliminating the oxygen content requirement for reformulated gasoline[9] and adding a provision directing the Administrator of the EPA to establish standards for toxic air pollutants from use of reformulated gasoline that maintain the reductions of annual aggregate emissions of such pollutants.[10]

[5] Loan Guarantees and Grants

Pursuant to Section 1510 of the Act, the DOE must establish a program to provide guarantees of loans by private institutions for the construction of facilities for the processing and conversion of municipal solid waste and cellulosic biomass into fuel ethanol and other commercial byproducts.[11] The authority to issue the guarantees terminates on August 8, 2015.[12]

In addition, Section 1511 of the Act directs the Secretary of Energy to issue loan guarantees to carry out no more than four projects to demonstrate, within certain parameters, the feasibility and viability of producing cellulosic biomass ethanol or sucrose-derived ethanol.[13]

---

[7] 2005 Energy Policy Act § 1502.

[8] 2005 Energy Policy Act § 1503.

[9] 2005 Energy Policy Act § 1504(a).

[10] 2005 Energy Policy Act § 1504(b).

[11] 2005 Energy Policy Act § 1510(b).

[12] 2005 Energy Policy Act § 1510(l).

[13] 2005 Energy Policy Act § 1511.

Under Section 1516 of the Act, the Secretary may also issue loan guarantees to projects to demonstrate the commercial feasibility and viability of producing ethanol using sugarcane, sugarcane bagasse, and other sugarcane byproducts as a feedstock.[14] Lastly, the Act authorizes the Secretary to provide conversion assistance to merchant producers of cellulosic biomass ethanol, waste-derived ethanol, and approved renewable fuels in the form of grants intended to assist such producers in building eligible production facilities.[15]

[6] Advanced Biofuel Technologies Program

Section 1514 of the Act directs the establishment of the Advanced Biofuel Technologies Program, with the purpose of demonstrating advanced technologies for the production of alternative transportation fuels. The Act further directs the funding of various demonstration projects under the program.[16]

[7] Studies and Analyses

The Act requires various agencies to conduct certain studies and analyses relative to the use and impact of renewable fuels. Section 1501(a) of the Act directs the FTC to perform an annual market concentration analysis of the ethanol production industry to assess the level of competition within the industry, in an effort to avoid price-setting and other anti-competitive behavior.[17]

Pursuant to Section 1501(d) of the Act, the Administrator of the EPA must conduct two surveys: (1) an annual survey to determine the market shares of conventional and reformulated gasoline containing ethanol or renewable fuel in each conventional gasoline use area and each reformulated gasoline use area in each state;[18] and (2) a monthly published survey of renewable fuels demand in the motor vehicle fuels market in the United States.[19]

In addition, Sections 1505 and 1509 of the Act direct the Administrator of the EPA to conduct two studies: (1) a study on the effects

---

[14] 2005 Energy Policy Act § 1516.

[15] 2005 Energy Policy Act § 1512.

[16] 2005 Energy Policy Act § 1514.

[17] 2005 Energy Policy Act § 1501(a).

[18] 2005 Energy Policy Act § 1501(d).

[19] 2005 Energy Policy Act § 1508.

on public health, air quality, and water resources of the increased use of, and feasibility of using as a substitute for MTBE, various fuel additives and blendstocks;[20] and (2) a study of federal, state, and local requirements regarding motor vehicle fuels, along with the Administrator's recommendations for legislative and administrative actions that may be taken to improve air quality, reduce costs to consumers and producers, and increase supply liquidity.[21]

Pursuant to Section 1506 of the Act, the Administrator must publish an analysis, within four years of enactment of the Act, of the changes in air pollutant emissions and air quality due to the use of motor vehicle fuel and fuel additives resulting from implementation of the Act. In addition, the Administrator must develop and finalize an emissions model that reflects the effects of gasoline characteristics and components on emissions from vehicles in the motor vehicle fleet in the 2007 calendar year.[22]

[8] Underground Storage Tank Compliance

Section 1522 of the Act amends Section 9004 of the Solid Waste Disposal Act by requiring the distribution of at least 80% of funds appropriated each fiscal year from the Leaking Underground Storage Tank Trust Fund. States may use the funds to pay for the reasonable costs of (1) corrective actions, (2) necessary administrative expenses directly related to state fund or state assurance programs, or (3) enforcement of state or local regulations relating to underground storage tanks.[23]

The Act also provides for the inspection of underground storage tanks, requiring inspections of all tanks regulated under the Act at least once every three years. With respect to regulated tanks that have not been inspected since December 22, 1998, the Act requires the initial inspection to occur within two years of the date of enactment.[24] In addition, the Act specifies training requirements for persons responsible for on-site operation and maintenance of underground storage tank systems.[25]

---

[20] 2005 Energy Policy Act § 1505.

[21] 2005 Energy Policy Act § 1509.

[22] 2005 Energy Policy Act § 1506.

[23] 2005 Energy Policy Act § 1522(a).

[24] 2005 Energy Policy Act § 1523(a).

[25] 2005 Energy Policy Act § 1524.

Section 1530 of the Act amends Section 9003 of the Solid Waste Disposal Act by adding a provision requiring the secondary containment and monitoring of any underground storage tank and related piping replaced or installed after the effective date of the subsection if the tank or piping is within 1,000 feet of any existing community water system.[26]

[9] Boutique Fuels

Under the Clean Air Act, states may require drivers in their jurisdiction to use a special blend of fuel (referred to as a "boutique fuel") to address unique aspects of their local air quality. Drivers in these regions face limited supplies during times of high gas demand. To limit the pain of such shortages, the Act allows the Administrator of the EPA to temporarily waive a control or prohibition regarding use of a fuel or fuel additive required by the Administrator or prescribed by a state implementation plan (SIP) if the Administrator determines that: (1) extreme or unusual fuel or fuel additive supply circumstances exist, (2) the circumstances are the result of a natural disaster or other unforeseen event, and (3) it is in the public interest to grant the waiver. The waiver must apply to the smallest geographic area necessary and is effective for a period of 20 calendar days or less.[27]

Also, to limit the proliferation of boutique fuels, the Act amends Section 211(c)(4)(C) of the Clean Air Act by adding a provision stating that the Administrator has no authority to approve any SIP or revision if such approval would increase the total number of fuels approved as of September 1, 2004, in all SIPs.[28]

Lastly, the Act directs the EPA, in coordination with the DOE, to conduct a study of the effects of the control and prohibition of certain fuels and fuel additives under SIP provisions adopted pursuant to Section 211(c)(4)(C) of the Clean Air Act. The goal of the study, which is to be submitted to Congress within 12 months of the date of enactment of the Act, is to determine how to develop a federal fuels system that maximizes motor fuel fungibility and supply, addresses air quality requirements, and reduces motor fuel price volatility.[29]

---

26 2005 Energy Policy Act § 1530(a).

27 2005 Energy Policy Act § 1541(a).

28 2005 Energy Policy Act § 1541(b).

29 2005 Energy Policy Act § 1541(c).

§ 1.16 Title XVI—Climate Change

[1] Scope

Title XVI does not impose any limits or caps on greenhouse gases. Instead, the primary focus of the Climate Change Title is to reduce greenhouse gas[1] both in the United States and abroad through research and development of greenhouse gas intensity reducing technologies. The Act also provides for the creation of a national strategy to promote technology to reduce greenhouse gases both domestically and globally. Congress created a new cabinet-level advisory committee to develop national policy regarding climate change and to promote technologies to reduce greenhouse gas emissions.

[2] Greenhouse Gas Intensity Reducing Strategies

Within 180 days of enactment of the Act, the President must establish a Committee on Climate Change Technology (Climate Change Committee) to integrate current federal climate reports, to coordinate federal climate change programs, and to develop a strategy to promote the deployment and commercialization of greenhouse gas intensity reducing technologies and practices.[2] This Climate Change Committee will be composed of at least seven members, including the Secretaries of Energy, Commerce, and Transportation.[3] The Climate Change Committee must submit its strategy to the Secretary of Energy and the President within 18 months of enactment of the Act. These research and development programs are to be conducted by the National Laboratories and federal research facilities, as well as institutions of higher education and the private sector. The Climate Change Committee must update the strategy at least every five years.[4]

Section 1610(d) of the Act requires the Secretary of Energy to establish a new Climate Change Technology Program (Program) within 180 days of establishing the Climate Change Committee. The Secretary is to establish the program in consultation with the Climate

---

[1] "Greenhouse gas" means carbon dioxide, methane, nitrous oxide, hydroflourocarbons, perflourocarbons, and sulfur hexaflouride. 2005 Energy Policy Act § 1610(a)(5).

[2] "Greenhouse gas intensity" means the ratio of greenhouse gas emissions to economic output. 2005 Energy Policy Act § 1610(a)(6).

[3] 2005 Energy Policy Act § 1610(b).

[4] 2005 Energy Policy Act § 1610(c)(2)(A)–(B).

Change Committee. The Program is to assist the Climate Change Committee in the interagency coordination of climate change technology research, development, demonstration, and deployment to reduce greenhouse gas intensity and to carry out the programs authorized.

The Secretary of Energy shall conduct an inventory and evaluation of greenhouse gas intensity reducing technologies that have been or are being developed to determine which technologies are suitable for commercialization or deployment. This inventory is to be made public and is to be followed by a report to Congress within 180 days of its completion. The inventory is to be used by the Secretary of Energy as guidance in the commercialization and deployment of greenhouse gas intensity-reducing technologies. The inventory is to be periodically updated.[5]

Section 1610 of the Act also requires the Secretary of Energy, in consultation with the Climate Change Committee, to establish a Climate Change and Technology Advisory Committee under the Department of Energy Organization Act (Advisory Committee).[6] This Advisory Committee would identify statutory, regulatory, economic, and other barriers to the commercialization and deployment of technologies for reducing greenhouse gas intensity. The Advisory Committee shall include a representative from each National Laboratory, representatives of energy-producing trade organizations, representatives of energy-intensive trade organizations, end-use energy and other consumers, federal government employees that are experts in energy technology, intellectual property, and tax, and representatives of higher education institutions.[7]

Within a year of enactment of the Act, and then annually, the Advisory Committee shall report to the Climate Change Committee any findings or recommendations for the removal or reduction of barriers to commercialization, deployment, and increasing the use of greenhouse gas intensity reducing technologies and practices.[8]

Based on the strategy and information from the reports created under this title, the Climate Change Committee is to develop recommendations to remove barriers to commercialization and deployment of

[5] 2005 Energy Policy Act § 1610(e).

[6] 42 U.S.C. § 7243.

[7] 2005 Energy Policy Act § 1610(f)(2)(A)–(F).

[8] 2005 Energy Policy Act § 1610(f)(3).

greenhouse gas intensity-reducing technologies and practices.[9] The Climate Change Committee is to consider cost-effectiveness, fiscal, regulatory, and statutory barriers, and intellectual property issues in developing the recommendations it makes.

The Climate Change Committee may also identify the need for climate change technology demonstration projects. Within 18 months of enactment of the Act, the Climate Change Committee is to submit to the President and Congress a report that identifies barriers to, and risks associated with, the deployment of greenhouse gas reducing technologies and that includes a plan for carrying out demonstration projects. The Climate Change Committee shall make the report available to the public and update the report at least every five years.[10]

Section 1610(h) of the Act also provides that, after public notice and comments, the Secretary of Energy, in collaboration with the Climate Change Committee and the National Institute of Standards and Technology, shall develop standards and best practices for calculating, monitoring, and analyzing greenhouse gas intensity.[11]

The Secretary of Energy shall, under Section 1610(i) of the Act, support demonstration projects that increase the reduction of greenhouse gas intensity to levels below that which would be achieved by technologies being used in the United States in 2005.[12]

Section 1610(j) of the Act authorizes the Secretary to enter into cooperative research and development agreements (CRADAs) under Section 12 of the Stevenson-Wydler Technology Innovation Act of 1980.[13] CRADAs between federal laboratories and private companies in the United States are intended, in large part, to transfer technologies developed at federal laboratories to private companies to facilitate commercialization of new technologies.

---

[9] 2005 Energy Policy Act § 1610(g).

[10] 2005 Energy Policy Act § 1610(g)(5).

[11] For current global emissions data, see Congressional Research Service, CRS Report RL32721, *Greenhouse Gas Emissions: Conflicting Situations, Conflicting Perspectives* (Jan. 2005).

[12] 2005 Energy Policy Act § 1610(i).

[13] 15 U.S.C. § 3710(a).

[3] Climate Change Technology Deployment in Developing Countries

Section 1611 of the Act adds a new Part C to the Global Environmental Protection Assistance Act of 1989 (GEPA) to task the Department of State as the lead agency for integrating into United States foreign policy the goal of reducing greenhouse gas intensity in developing countries.[14] The Secretary of State must submit an initial report to the appropriate congressional committees within 180 days of enactment of the Act that identifies the 25 developing countries that are the largest emitters of greenhouse gas. For each of these 25 countries, the report must include an estimate of the quantity and types of energy used and of the greenhouse gas intensity of the nation's energy, manufacturing, agricultural, and transportation sectors. The report must also include descriptions of the progress of projects undertaken to reduce greenhouse gas intensity, potential for undertaking new projects to reduce greenhouse gas intensity, obstacles to the reduction of greenhouse gas intensity, and best practices learned by the United States Agency for International Development (U.S. AID) from conducting previous pilot and demonstration projects.[15] This report must be updated within 18 months of enactment of the Act to track the progress of the developing countries.[16]

The Secretary of State and the U.S. AID Administrator must assist developing countries to reduce greenhouse gas intensity, including fostering private investment in projects and expedited development of technology.[17] This assistance can be provided directly or through agreements with the World Bank, International Monetary Fund, Overseas Private Investment Corporation, and other development institutions. In providing this assistance, the Secretary of State is to focus on promoting the rule of law, property rights, contract protection, economic freedom, increasing capacity, infrastructure, and training.

[14] 2005 Energy Policy Act § 1611; GEPA, Pub. L. No. 101-240; 103 Stat. 2521 (1989) § 732(a).

[15] 2005 Energy Policy Act § 732; GEPA § 732.

[16] Section 732(a)(2)(B) of the Act refers to an "update" of information provided in the initial report, but § 732(a)(2)(C) refers to "annual reports." However, there is no indication in the initial reference to "updates" that the report is to subsequently or annually updated.

[17] 2005 Energy Policy Act § 1611; GEPA § 733.

The Act requires Secretary of State to give priority to projects in the 25 developing countries identified in the report submitted under Section 732(a)(2)(A) of the GEPA.[18]

[4] Projects and Technology Inventory in Developing Countries

The Secretary of Energy, in consultation with the Secretaries of State and Commerce, shall conduct an inventory of greenhouse gas intensity reducing technologies under development in the United States to identify technologies suitable for export to developing countries.

The Secretary of State, in coordination with the Secretaries of Energy and Commerce, must submit to Congress a report that identifies the technologies; identifies obstacles to the transfer, deployment, and commercialization of the inventoried technologies; identifies the results from previous federal reports related to the inventoried technologies; and provides an analysis of market forces related to the inventoried technologies.[19]

[5] Trade-Related Barriers to Export of Greenhouse Gas Intensity Reducing Technologies

By August 8, 2006, the U.S. Trade Representative (USTR) shall identify trade barriers maintained by foreign countries that interfere with the export of greenhouse gas intensity reducing technologies and practices from the United States to the 25 developing countries identified in the report submitted to Congress. The USTR is to negotiate with foreign countries for the removal of those barriers as is appropriate and consistent with applicable bilateral, regional, and mutual trade agreements. The USTR shall, within one year after the report is submitted under Section 734(a)(1) of the GEPA, and annually thereafter, submit to Congress a report that describes any progress made with respect to removing the barriers identified.

---

[18] The focus on assistance for developing countries is most likely in response to the fact that developing countries are exempt from the obligations of the Kyoto Protocol yet are projected to account for two thirds of the growth in carbon dioxide emissions in the next 20 years. *See* Congressional Budget Office Report, *Uncertainty in Analyzing Climate Change: Policy Implications* (Jan. 2005).

[19] 2005 Energy Policy Act § 1611; GEPA § 733(b).

[6] Greenhouse Gas Intensity Reducing Technology Export Initiative

New Section 735(a) of the GEPA establishes the Greenhouse Gas Intensity Reducing Technology Export Initiative, which is an interagency working group to promote the export of technologies and practices from the United States and to identify developing countries that should be designated as priority countries for the purpose of exporting such technologies and practices based on the report submitted under Section 732(a)(2)(A) of the GEPA. The group is also to identify potential barriers that prevent adoption of exported greenhouse gas intensity-reducing technologies and practices and to identify previous efforts to export energy technologies to learn best practices. The working group shall, within 180 days of enactment of the Act, and annually thereafter, conduct performance reviews and submit reports and recommendations to appropriate congressional committees.[20]

The Secretaries of State and Energy and the U.S. AID Administrator shall plan, coordinate, and carry out (or provide assistance for) demonstration projects in at least 10 eligible countries.[21] A country shall be eligible for assistance if the Secretaries and U.S. AID Administrator determine that the country has demonstrated a commitment to (1) just governance, including promoting the rule of law, respecting human and civil rights, protecting private property rights and combating corruption; and (2) economic freedom, including economic policies that encourage citizens and firms to participate in global trade and international capital markets, promote private-sector growth and the sustainable management of natural resources, and strengthen market forces in the economy.

§ 1.17 Title XVII—Incentives for Innovative Technologies

[1] Scope

Congress added Title XVII to the Act to facilitate development of innovative technologies that would reduce greenhouse gases or employ new or significantly improved technologies for the production of energy sources. Under the Title XVII, Congress created a loan

---

20 2005 Energy Policy Act § 735(c).

21 2005 Energy Policy Act § 736.

guarantee program to provide financial encouragement for such activities. The program is to be administered by the Secretary of Energy.

[2] The Loan Program

Section 1701 of the Act sets forth the technical definitions to be used in the loan guarantee program. The exact meanings of "cost" and "guarantee" are those set forth in Section 502 of the Federal Credit Reform Act of 1990.[1] Section 1702 of the Act lays out the general terms, conditions, and limitations of the program.

For a particular loan to receive the benefit of a guarantee under the program, there must first be an appropriation for the guarantee or the Secretary must have received from the borrower a payment in full of the "cost" of the obligation.[2] In addition, the Secretary has been authorized to assess fees to recover the expense of administering the guarantee program.[3] The amount of any loan being guaranteed is limited to 80% of projected cost of the facility.[4] Before granting the guarantee to which the full faith and credit of the United States has been pledged,[5] the Secretary must be satisfied that there are sufficient funds, both guaranteed and otherwise, to complete the project and that there is a reasonable prospect of repayment of both principal and interest, with such obligations not being subordinate to any other financing.[6] When setting the rate of interest, the Secretary must take into consideration the prevailing rate of interest in the private sector for similar loans and risks.[7] Repayment of the loan must be fulfilled in the lesser of 30 years or 40% of the projected useful life of the physical asset financial by the obligation.[8]

---

1 2 U.S.C. § 661a.

2 2005 Energy Policy Act § 1702(b). When crafting the conditions of a guarantee, the Secretary of Energy is required to consult with the Secretary of the Treasury. 2005 Energy Policy Act § 1702(a).

3 2005 Energy Policy Act § 1702(h).

4 2005 Energy Policy Act § 1702(c).

5 2005 Energy Policy Act § 1702(j).

6 2005 Energy Policy Act § 1702(d).

7 2005 Energy Policy Act § 1702(e).

8 2005 Energy Policy Act § 1702(f).

In the event of a default, by the borrower, the holder of the guarantee has the right to demand, on the terms and conditions specified in the guarantee or related agreements, the balance of principal and unpaid interest from the Secretary who must comply, unless the Secretary finds that there was no default by the borrower on the principal or interest or that any default has been remedied.[9] The holder of the guarantee may forebear demanding payment, which may be agreed upon by the parties to the obligation and approved by the Secretary.[10] Where the federal government has made a payment on the default by the borrower, the Secretary becomes subrogated to the rights of the recipient of payment, including the authority to complete, maintain, operate, lease, or otherwise dispose of the project or permit the borrower to continue with the project.[11] The Secretary's rights are superior to the rights of any other person.[12] The statute also requires the Secretary, as part of any guarantee, to protect the interests of the United States and have available all the patents and technology necessary for any person, including the Secretary, to complete and operate the project.[13] To provide some flexibility outside of a formal default, the Secretary, for and on behalf of a borrower, may contract to pay holders of the obligation, the principal and interest which become due and payable on the unpaid balance where the Secretary finds that the borrower cannot pay, the public interest would benefit from the continuance of the project, there is probable net benefit to the federal government greater than a payment in the event of a default, the payment would be greater than the default payment, and the borrower has agreed to reimburse the Secretary for the payment, including interest.[14]

In the event of a default, the Secretary must notify the Attorney General who is required to take appropriate action to recover the unpaid principal and interest from the assets underlying the loan or from any other security pledged to serve the obligation.[15]

---

[9] 2005 Energy Policy Act § 1702(g).

[10] 2005 Energy Policy Act § 1702(g)(1)(C).

[11] 2005 Energy Policy Act § 1702(g)(2)(A).

[12] 2005 Energy Policy Act § 1702(g)(2)(B).

[13] 2005 Energy Policy Act § 1702(g)(2)(C).

[14] 2005 Energy Policy Act § 1702(g)(3).

[15] 2005 Energy Policy Act § 1702(g)(4).

For the purpose of keeping track of the financial status of the projects enjoying the benefit of the guarantee program and thereby facilitating effective audits by the Secretary or the Comptroller General of the United States, all recipients of guarantees are required to keep records in a manner prescribed by the Secretary.[16]

[3] Eligible Facilities

The Act specifies that the loan guarantees should be for projects that avoid, reduce, or sequester air pollutants or anthropogenic emissions of greenhouse gasses and employ new or improved technology to achieve the reduction. The facilities that may qualify for the guarantees include (1) renewable energy systems, (2) advanced fossil energy (including qualifying coal gasification), (3) hydrogen fuel cell technology for residential, industrial, or transportation uses, (4) advanced nuclear energy facilities, (5) carbon capture and sequestration practices and technologies, including agricultural and forestry practices that store and requestor carbon, (6) efficient electrical generation, transmission, and distribution technologies, (7) efficient end-use energy technologies, (8) production facilities for fuel efficient vehicles, including hybrid and advanced diesel vehicles, (9) pollution control equipment, and (10) refineries for converting crude oil into gasoline.[17]

The Act establishes specific criteria for loan guarantees for gasification projects. Congress divided such projects into four groups: (1) integrated gasification combined cycle projects, (2) industrial gasification projects, (3) petroleum coke gasification projects, and (4) liquefaction projects.

Within the first of these groups, integrated gasification combined cycle projects, Congress set forth five different categories of precise qualifying standards. Where the generation of electricity is involved, standards vary depending on the nature of the fossil fuel and the location of the fuel source. The first of the five categories of standards applies to generating facilities for which, during the term of the guarantee, coal, biomass, petroleum coke, or a combination thereof will account for at least 65% of the annual heat input and the electricity

[16] 2005 Energy Policy Act § 1702(i).

[17] 2005 Energy Policy Act § 1703(b)(3).

produced will account for at least 65% of annual heat output. Construction for such a generating facility must commence within three years after the issuance of the guarantee and must be of a design that, as determined by the Secretary, is capable of accommodating equipment likely to be necessary to capture the carbon dioxide that would otherwise be emitted as flue gas. The guarantee is also dependent on there being an assured revenue stream that covers the project's capital debt service, and a level of operating costs that is approved by the Secretary and the relevant state public utility commission.

The second of the five classes of integrated gasification combined cycle projects eligible for a guarantee are those whose energy source comes from coal having not more than 13,000 Btu/lb and mined in the western United States. These projects must rely upon appropriate advanced integrated gasification combined cycle technology that minimizes and offers the potential to sequester carbon dioxide emissions. These projects may include the repowering of existing facilities. They also may be built in stages. However, such projects must have a combined output of at least 100 megawatts, must be located in a western state at an altitude greater than 4,000 feet, and must demonstrate the ability to use coal with an energy content of not more than 9,000 Btu/lb.

The third group of qualifying integrated gasification combined cycle projects consists of those projects located in a taconite-producing region of the United States that are entitled by the law of the state in which the plant is located to enter into a long-term contract approved by a state public utility commission to sell at least 450 megawatts of output to a utility.

The fourth group of integrated gasification combined cycle projects consists of those facilities that generate one or more hydrogen-rich and carbon monoxide-rich product streams from the gasification of coal or coal waste and use those streams to facilitate the production of ultra clean premium fuels through the Fischer-Tropsch process.

The last qualifying group of eligible integrated gasification combined cycle projects includes facilities producing energy and clean fuels, using appropriate coal liquefaction technology, from "Western" bituminous or sub-bituminous coal that is owned by a state government, to which may be added tribal and private coal resources.

Beyond combined cycle facilities, the second of the four classes of gasification projects eligible for loan guarantees consists of industrial facilities that gasify coal, biomass, or petroleum coke in any combination to produce synthesis gas for use as a fuel or feed stock and for which electricity accounts for less than 65% of useful energy output of the facility.

As noted, the second class of eligible gasification projects includes certain facilities that gasify petroleum coke. However, Congress took the further step of identifying petroleum coke gasification projects as a category unto itself—the third category of loan-guarantee-eligible gasification projects. Moreover, Congress expressly "encouraged" the Secretary to make loan guarantees available for such projects.

The fourth and final group of eligible gasification projects is coal-to-oil liquefaction projects. For such projects, Congress tied this loan guarantee program in Title XVII to the funds available for the Clean Coal Power Initiative under Title IV. Funds awarded under that Title IV program for coal-to-oil liquefaction projects are expressly permitted to finance the costs of loan guarantees for such projects under Title XVII.

All projects that qualify for the loan guarantees must meet certain emission standards: the total sulfur dioxide emission in the flue gas may not exceed 0.05 lb/MMBtu; the project must have a 90% removal rate (including any fuel pretreatment) of mercury from the coal-derived gas and any other fuel combusted by the project; total nitrogen oxide emissions in the project's flue gas may not exceed 0.08 lb/MMBtu; and total particulate emissions in the flue gas are limited to no more than 0.01 lb/MMBtu.[18]

The fact that a particular project receives a tax credit for use of a clean coal technology does not disqualify it from receiving a loan guarantee.[19]

In Section 1704 of the Act, Congress authorized to be appropriated the necessary sums to provide the cost of the guarantees.[20] In addition, Congress has permitted the DOE to "use amounts awarded under the clean coal power initiative under subtitle A of title IV to carry out

[18] 2005 Energy Policy Act § 1703(d).

[19] 2005 Energy Policy Act § 1703(e).

[20] 2005 Energy Policy Act § 1703(e).

the project described in section 1703(c)(1)(C), on the request of the recipient of such award, for a loan guarantee, to the extent that the amounts have not yet been disbursed to, or have been repaid by, the recipient."[21]

§ 1.18 Title XVIII—Studies

[1] Scope

Title XVIII requires various federal agencies to perform 40 separate studies, investigations, or analyses to address a variety of issues. Generally, the responsible agency is required to submit reports to the Congress and/or the President. Each section in Title XVIII sets forth the responsible agency, detailed study requirements, and the timeframe of the study.

[2] Study on Inventory of Petroleum and Natural Gas Storage

Storage capability plays an increasingly important role in the marketplace and in the hydrocarbon industry's ability to meet demand. Likely because the relationship between storage capacity and market price could be significant in the current context of oil and natural gas markets, Congress required the Secretary of Energy to conduct a study on petroleum and natural gas storage capacity and operational inventory levels, nationwide and by major geographical regions.[1] The study must address: (1) historical normal ranges for petroleum and natural gas inventory levels; (2) historical and projected storage capacity trends; (3) estimated operation inventory levels below which outages, delivery slowdown, rationing, interruptions in service, or other indicators of shortage begin to appear; (4) explanations for inventory levels dropping below normal ranges; and (5) the ability of industry to meet United States demand for petroleum and natural gas without shortages or price spikes, when inventory levels are below normal ranges. The Secretary of Energy must report the study results, including findings

---

[21] 2005 Energy Policy Act § 1704(b). The reference in the text to Section 1703(c)(1)(C) probably should have been to Section 1703(e)(4) which discusses loan guarantees for projects coming under the clean coal power initiative.

[1] For purposes of this study, the term "petroleum" is defined to mean "crude oil, motor gasoline, jet fuel, distillates, and propane." 2005 Energy Policy Act § 1801(a).

and any recommendations for preventing future supply shortages, to Congress no later than one year after August 8, 2005.[2]

[3] Study of Energy Efficiency Standards

Section 1802 of the Act requires the Secretary of Energy to contract with the National Academy of Sciences (NAS) to study whether the goals of energy efficiency standards are best served by measurement of energy consumed, and efficiency improvements, at the actual site of energy consumption, or through the full fuel cycle, beginning at the source of energy production. The study must be completed no later than August 8, 2006. The Secretary of Energy is required to submit the report to Congress.[3]

[4] Telecommuting Study

The Secretary of Energy, in consultation with the Federal Communications Commission, the Director of the Office of Personnel Management, the Administrator of General Services, and the Administrator of National Telecommunications and Information Administration of the Department of Commerce, must conduct a study of the energy conservation implications of the widespread adoption of telecommuting by federal employees in the United States.[4] The Secretary of Energy must submit a report on the study to the President and Congress no later than February 8, 2006.[5]

[5] LIHEAP Report

The Secretary of Health and Human Services must report to Congress on how the Low-Income Home Energy Assistance Program (LIHEAP) could be used more effectively to prevent loss of life from extreme temperatures. The Secretary of Health and Human Services must consult with appropriate officials in all 50 states and the District of Columbia. The report must be made no later than August 8, 2006.[6]

---

[2] 2005 Energy Policy Act § 1801.

[3] 2005 Energy Policy Act § 1802.

[4] The term "telecommuting" means the performance of work functions using communications technologies, thereby eliminating or substantially reducing the need to commute to and from traditional worksites. The term "Federal employee" has the meaning provided the term "employee" by Section 2105 of Title 5, United States Code. 2005 Energy Policy Act § 1803(d)(3)–(4).

[5] 2005 Energy Policy Act § 1803.

[6] 2005 Energy Policy Act § 1804.

[6] Oil Bypass Filtration Technology

Section 1805 of the Act requires the Secretary of Energy and the Administrator of the EPA to conduct a joint study of the benefits of oil bypass filtration technology in reducing demand for oil and protecting the environment and to examine the feasibility of using oil bypass filtration technology in federal motor vehicle fleets. The study must evaluate products and various manufacturers before determining the feasibility of using oil bypass filtration technology. Section 1805 provides no deadline for completion.[7]

[7] Total Integrated Thermal Systems

The Secretary of Energy must study the benefits of total integrated thermal systems in reducing demand for oil and protecting the environment. The Secretary of Energy also must examine the feasibility of using total integrated thermal systems in Department of Defense and other federal motor vehicle fleets. Section 1806 of the Act provides no deadline for completion.[8]

[8] Report on Energy Integration With Latin America

Section 1807 of the Act requires the Secretary of Energy to submit an annual report to the Committee on Energy and Commerce of the United States House of Representatives and to the Committee on Energy and Natural Resources of the Senate concerning the status of energy export development in Latin America and the efforts of the Secretary of Energy and other departments and agencies of the United States to promote energy integration with Latin America.

The annual reports must contain a detailed analysis of the status of energy export development in Mexico and a description of all significant efforts by the Secretary of Energy and other departments and agencies to promote a constructive relationship with Mexico regarding the development of that nation's energy capacity. More specifically, the annual reports must outline efforts the Secretary of Energy and other departments and agencies have made to ensure that regulatory approval and oversight of United States/Mexico border projects that result in the expansion of Mexican energy capacity are

[7] 2005 Energy Policy Act § 1805.

[8] 2005 Energy Policy Act § 1806.

effectively coordinated across departments and with the Mexican government.[9]

[9] Low-Volume Gas Reservoir Study

The Secretary of Energy must make a grant to an organization of oil and gas producing states with significant numbers of marginal oil and natural gas wells that will permit the organization to conduct annual studies of low-volume natural gas reservoirs.[10]

The annual studies required by Section 1808 of the Act must: (1) determine the status and location of marginal wells and gas reservoirs; (2) gather the production information of these marginal wells and reservoirs; (3) estimate the remaining producible reserves based on variable pipeline pressures; (4) locate low-pressure gathering facilities and pipelines; (5) recommend incentives which will enable the continued production of these resources; (6) produce maps and literature to disseminate to states to promote conservation of natural gas reserves; and (7) evaluate the amount of natural gas that is being wasted through the practice of venting or flaring of natural gas produced in association with crude oil well production.[11]

The Act authorizes the appropriation of the following funds to the Secretary of Energy for carrying out this section: $1,500,000 for fiscal year 2006 and $450,000 for each of the fiscal years 2007-2010.[12]

[10] Investigation of Gasoline Prices

The Federal Trade Commission (FTC) must conduct an investigation to determine whether the price of gasoline is being artificially manipulated by reduced refinery capacity or by any other form of market manipulation or price gouging practices. This investigation must be conducted no later than 90 days after November 8, 2005. Upon completion of the FTC's investigation, the FTC is required to submit to Congress a report that describes the results of the investigation and any FTC recommendations.

Section 1809 of the Act also requires the Secretary of Energy to direct the National Petroleum Council (NPC) to evaluate whether, and

[9] 2005 Energy Policy Act § 1807.

[10] This organization must work with the state geologist of each state being studied.

[11] 2005 Energy Policy Act § 1808.

[12] 2005 Energy Policy Act § 1808.

to what extent, environmental and other regulations affect new domestic refinery construction and significant expansion of existing refinery capacity. Upon completion of the NPC's evaluation and analysis, the Secretary of Energy must submit a report to Congress that describes the results of the evaluation and analysis, and any NPC recommendations.[13]

[11] Alaska Natural Gas Pipeline

Section 1810 of the Act requires the Federal Energy Regulatory Commission (FERC) to submit to Congress a report describing the progress made in licensing and constructing the pipeline, and any issue impeding that progress. The first report must be made no later than 180 days after enactment.[14] Additional reports must be submitted every 180 days thereafter until the Alaska natural gas pipeline commences operation.[15]

[12] Coal Bed Methane Study

Section 1811 of the Act requires the Secretary of the Interior, in consultation with the Administrator of the EPA, to enter into an arrangement under which the NAS will conduct a study on the effect of coal bed natural gas production on surface and ground water resources, including ground water aquifers, in the States of Montana, Wyoming, Colorado, New Mexico, North Dakota, and Utah.

The study must address the effectiveness of the management of coal bed methane produced water, the use of best management practices, and various production techniques for coal bed methane natural gas in minimizing impacts on water resources. The study is required to analyze available hydrologic, geologic and water quality data, along with: (1) production techniques, produced water management techniques, best management practices, and other factors that can mitigate effects of coal bed methane development; (2) the costs associated with mitigation techniques; (3) effects on surface or ground water resources, including drinking water, associated with surface or subsurface

[13] 2005 Energy Policy Act § 1809.

[14] As of November 23, 2005, the FERC had not initiated its process to produce these reports. See the FERC website at www.ferc.gov/legal/maj-ord-reg/fed-sta/ene-pol-act.asp.

[15] 2005 Energy Policy Act § 1810.

disposal of waters produced during extraction of coal bed methane; and (4) any other significant effects on surface or ground water resources associated with production of coal bed methane.

The study must also analyze the effectiveness of current mitigation practices for handling coal bed methane produced water in relation to existing federal and state laws and regulations, and make recommendations as to changes, if any, to federal law necessary to address adverse impacts to surface or ground water resources associated with coal bed methane development.[16]

Section 1811(d) of the Act requires the NAS to submit the findings and recommendations of the study to the Secretary of the Interior and the Administrator of the EPA no later than August 8, 2006. The NAS also must make study results available to the public. No later than six months after receiving the study results, the Secretary of the Interior and the Administrator of the EPA, after consulting with states, must report to the Congress on the study's findings and any recommendations in funding to address the effects of coal bed methane production on surface and ground water resources.[17]

[13] Backup Fuel Capability Study

The Secretary of Energy must conduct a study of the effect of obtaining and maintaining liquid and other fuel backup capability at gas-fired power generation facilities, and other gas-fired industrial facilities. The study must address the costs and benefits of adding a different fuel capability to a power gas-fired power generating or industrial facility, taking into consideration regional differences. The study also must address methods that the federal and state governments can use to encourage gas-fired power generators and industries to develop the capability to power the facilities using a backup fuel. Further, the study must address the effect on the supply and cost of natural gas of a balanced portfolio of fuel choices in power generation and industrial applications and state regulations that permit agencies in the state to carry out policies that encourage the use of other backup fuels in gas-fired power generation. Finally, the study must address

[16] 2005 Energy Policy Act § 1811.

[17] 2005 Energy Policy Act § 1811.

changes required in the Clean Air Act[18] to allow natural gas generators to add clean backup fuel capabilities.[19]

The Secretary of Energy must report the results of the study to Congress. The report must include recommendations regarding future activity of the federal government relating to backup fuel capability. The report must be submitted no later than August 8, 2006.

[14] Indian Land Rights-of-Way

The Secretary of Energy and the Secretary of the Interior must jointly conduct a study of issues regarding energy rights-of-way on tribal land.[20] The report must include: (1) an analysis of historic rates of compensation paid for energy rights-of-way on tribal land; (2) recommendations for appropriate standards and procedures for determining fair and appropriate compensation to Indian tribes for grants, expansions, and renewals of energy rights-of-way on tribal land; (3) an assessment of the tribal self-determination and sovereignty interests implicated by applications for the grant, expansion, or renewal of energy rights-of-way on tribal land; and (4) an analysis of relevant national energy transportation policies relating to grants, expansions, and renewals of energy rights-of-way on tribal land.[21] In conducting this joint study the Secretaries are required to consult with Indian tribes, the energy industry, appropriate governmental entities, and affected businesses and consumers. The Secretaries must report their findings to Congress no later than August 8, 2006.

[15] Mobility of Scientific and Technical Personnel

The Secretary of Energy must transmit to Congress a report identifying any policies or procedures of a contractor operating a National Laboratory or single-purpose research facility that create disincentives to the temporary or permanent transfer of scientific and technical personnel among the contractor-operated National Laboratories or contractor-operated single-purpose research facilities. The

[18] 42 U.S.C. § 7401 *et seq.*

[19] 2005 Energy Policy Act § 1812.

[20] The term "tribal land," in this circumstance, is defined in Section 2601 of the Energy Policy Act of 1992, as amended by Section 503 of the 2005 Energy Policy Act. 2005 Energy Policy Act § 1813(a)(1).

[21] 2005 Energy Policy Act § 1813.

report also must provide recommendations for improving interlaboratory exchange of scientific and technical personnel. The Secretary of Energy must transmit this report to Congress no later than two years after August 8, 2005.[22]

[16] Interagency Review of Competition in Wholesale and Retail Markets for Electric Energy

Section 1815 of the Act establishes the "Electric Energy Market Competition Task Force." This inter-agency task force consists of five members—one employee from each of the following: (1) the Department of Justice (appointed by the Attorney General of the United States); (2) the FERC (appointed by the Chairperson of the FERC); (3) the FTC (appointed by the Chairperson of the FTC); (4) the Department of Energy (DOE) (appointed by the Secretary of Energy); and (5) the Rural Utilities Service (RUS) (appointed by the Secretary of Agriculture).[23]

The joint Task Force is required to study and analyze competition in the wholesale and retail markets for electric energy.[24] No later than August 8, 2006, the joint Task Force must submit its findings to Congress. No later than 60 days before a final report is submitted to Congress, the joint Task Force must publish a draft of the report in the Federal Register, and provide an opportunity for public comment on the report. The Task Force must consult with, and solicit comments from, any advisory entity of the Task Force, the states, representatives of the electric power industry, and the public.[25]

---

22 2005 Energy Policy Act § 1814.

23 The initial members of the joint task force are: DOJ — J. Bruce McDonald, Deputy Assistant Attorney General, Antitrust Division; FERC — Michael Bardee, Associate General Counsel, Office of the General Counsel—Markets, Tariffs, and Rates; FTC — Michael Wroblewski, Assistant General Counsel for Policy Studies; DOE — David Meyer, Deputy Director, Division of Permitting, Siting, and Analysis, Office of Electricity Delivery and Energy Reliability; RUS — Karen Larsen, Office of Assistant Administrator, Electric Programs.

24 On October 13, 2005, FERC opened Docket No. AD05-17-000 and sought public comment on a series of questions related to competition in the retail and wholesale electric market.

25 2005 Energy Policy Act § 1815.

[17] Study of Rapid Electrical Grid Restoration

The Secretary of Energy must conduct a study of the benefits of using mobile transformers and mobile substations to rapidly restore electrical service to areas subjected to blackouts as a result of equipment failure, natural disasters, and acts of terrorism or war.

The study must contain an analysis of the feasibility of using mobile transformers and mobile substations to reduce dependence on foreign entities for key elements of the electrical grid system of the United States. It also must contain an analysis of the feasibility of using mobile transformers and mobile substations to rapidly restore electrical power to military bases, the federal government, communications industries, first responders, and other critical infrastructures, as determined by the Secretary of Energy. Finally, the study must include an analysis of the quantity of mobile transformers and mobile substations necessary to (1) eliminate dependence on foreign sources for key electrical grid components in the United States, (2) rapidly deploy technology to fully restore full electrical service to prioritized governmental functions, and (3) identify manufacturing sources in existence on August 8, 2005 that have previously manufactured specialized mobile transformer or mobile substation products for federal agencies.

No later than August 8, 2006, the Secretary of Energy must report the findings of this study to the President and Congress.[26]

[18] Study of Distributed Generation

The Secretary of Energy, in consultation with the FERC, must conduct a study of the potential benefits of cogeneration and small power production.[27]

The study must include an analysis of the potential benefits of: (1) increased system reliability; (2) improved power quality; (3) the provision of ancillary services; (4) reduction of peak power requirements through onsite generation; (5) the provision of reactive power or volt-ampere reactives; (6) an emergency supply of power; (7) offsets to investments in generation, transmission, or distribution facilities that

---

[26] 2005 Energy Policy Act § 1816.

[27] As of November 23, 2005, the FERC had not initiated its consultation with the Secretary of Energy. See the FERC website at www.ferc.gov/legal/maj-ord-reg/fed-sta/ene-pol-act.asp.

would otherwise be recovered through rates; (8) diminished land use effects and right-of-way acquisition costs; and (9) reducing the vulnerability of a system to terrorism.

The study also must include an analysis of any rate-related issue that may impede or otherwise discourage the expansion of cogeneration and small power production facilities, including a review of whether rates, rules, or other requirements imposed on the facilities are comparable to rates imposed on customers of the same class that do not have cogeneration or small power production.

No later than 18 months after enactment of the Act, the Secretary of Energy must complete the study, provide an opportunity for public comment on the results of the study, and submit a report on the study to the President and Congress. The report must describe the results of the study and the public comments. The Secretary of Energy must publish the report after it is submitted to the President and Congress.[28]

[19] Natural Gas Supply Shortage Report

The Secretary of Energy must submit to Congress a report on natural gas supplies and demand. The purpose of this report is to develop recommendations to achieve a balance between natural gas supply and demand in order to: (1) provide residential consumers with natural gas at reasonable and stable prices; (2) accommodate long-term maintenance and growth of domestic natural gas-dependent industrial, manufacturing, and commercial enterprises; (3) facilitate the attainment of national ambient air quality standards under the Clean Air Act;[29] (4) achieve continued progress in reducing the emissions associated with electric power generation; and (5) support the development of the preliminary phases of hydrogen-based energy technologies.[30]

The analysis must include recommendations for federal actions to (1) encourage or require the use of energy sources other than natural gas, including coal, nuclear, and renewable sources; (2) encourage or require the use of energy conservation or demand side management practices; (3) support technologies for the development of alternative

---

28 2005 Energy Policy Act § 1817.

29 43 U.S.C. § 7401 *et seq.*

30 2005 Energy Policy Act § 1817.

sources of natural gas and synthetic gas, including coal gasification technologies; and (4) improve access to domestic natural gas supplies.

In preparing the report, the Secretary of Energy is required to consult with experts in natural gas supply and demand, state and local government representatives, tribal organizations, and consumer and other organizations. If the Secretary of Energy considers it appropriate, the Secretary of Energy may hold public hearings and provide other opportunities for public comment. The report must be submitted to Congress no later than 180 days after August 8, 2005.[31]

[20] Hydrogen Participation Study

The Secretary of Energy must submit to Congress a report evaluating methodologies to ensure the widest participation practicable in setting goals and milestones under the hydrogen program of the DOE, including international participants. This report must be submitted to Congress no later than August 8, 2006.[32]

[21] Overall Employment in a Hydrogen Economy

The Secretary of Energy must study the likely effects of a transition to a hydrogen economy on overall employment in the United States. In completing the study, the Secretary of Energy must consider the replacement effects of new goods and services, international competition, workforce training requirements, multiple possible fuel cycles, including usage of raw materials, rates of market penetration of technologies, and regional variations based on geography. The Secretary of Energy must submit to Congress a report describing the findings, conclusions, and recommendations of such a study. The report is due no later than 18 months after August 8, 2005.[33]

[22] Study of Best Management Practices for Energy Research and Development Programs

The Secretary of Energy must enter into an arrangement with the National Academy of Public Administration under which the Academy will conduct a study to assess management practices for research, development, and demonstration programs at the DOE. The study must

[31] 2005 Energy Policy Act § 1818.

[32] 2005 Energy Policy Act § 1819.

[33] 2005 Energy Policy Act § 1820.

consider, among other things, management practices that act as barriers between the Office of Science and offices conducting mission-oriented research; and recommendations for management practices that would improve coordination and bridge the innovation gap between the Office of Science and offices conducting mission-oriented research. The Secretary of Energy must submit to Congress a report on the study no later than 18 months after August 8, 2005.[34]

[23] Effect of Electrical Contaminants on Reliability of Energy Production Systems

The Secretary of Energy must enter into a contract with the NAS pursuant to which the NAS will determine the effect that electrical contaminants (such as tin whiskers) may have on the reliability of energy production systems, including nuclear energy. This contract must be entered no later than 180 days after August 8, 2005.[35]

[24] Alternative Fuels Reports

The Secretary of Energy must submit to Congress reports on the potential for each of biodiesel and hythane to become major, sustainable, alternative fuels. Both reports must submitted no later than August 8, 2006.

The biodiesel report must provide detailed assessments of the potential for biodiesel becoming a substantial source of fuel for conventional diesel and heating oil applications, biodiesel markets and manufacturing capacity and the environmental, and the energy security benefits associated with the use of biodiesel. The hythane report must provide a detailed assessment of potential hythane markets and the research and development activities that are necessary to facilitate the commercialization of hythane as a competitive, environmentally friendly transportation fuel.[36]

[25] Final Action on Refunds for Excessive Charges

Section 1824 of the Act encourages the FERC to conclude its investigation into the unjust or unreasonable charges incurred by California during the 2000-2001 electricity crisis as soon as possible.

---

[34] 2005 Energy Policy Act § 1821.

[35] 2005 Energy Policy Act § 1822.

[36] 2005 Energy Policy Act § 1823.

The FERC also must seek to ensure that refunds the FERC determines are owed to the State of California are paid to the State of California. Finally, the FERC must submit to Congress a report by December 31, 2005, describing the actions taken by the Commission to date under this section and timetables for further actions.[37]

[26] Fuel Cell and Hydrogen Technology Study

As soon as practicable, the Secretary of Energy must enter into a contract with the NAS and the National Research Council (NRC) to study fuel cell technologies that will provide a budget roadmap for the development of fuel cell technologies and the transition from petroleum to hydrogen in a significant percentage of the vehicles sold by 2020.[38]

[27] Passive Solar Technologies

The Secretary of Energy must conduct a study to determine the range of levelized costs of avoided electricity for passive solar technologies.[39] The Secretary of Energy must submit its results to Congress no later than February 8, 2006.[40]

[28] Study of Link Between Energy Security and Increases in Vehicle Miles Traveled

The Secretary of Energy must enter into an arrangement with the NAS under which the NAS will conduct a study to assess the implications on energy use and efficiency of land development patterns in the United States. The study must consider the correlation, if any, between land development patterns and increases in vehicle miles traveled and whether petroleum use in the transportation sector can be reduced through changes in the design of development patterns. The NAS also must consider such other matters relating to the study topic as the NAS finds appropriate. The NAS must submit to the

[37] 2005 Energy Policy Act § 1824.

[38] 2005 Energy Policy Act § 1825.

[39] In this section, the term "passive solar technology" means a passive solar technology, including daylighting, that is used exclusively to avoid electricity use, and can be metered to determine energy savings. 2005 Energy Policy Act § 1826(a).

[40] 2005 Energy Policy Act § 1826.

Secretary of Energy and Congress a report on this study no later than two years after August 8, 2005.[41]

[29] Science Study on Cumulative Impacts of Multiple Offshore Liquefied Natural Gas (LNG) Facilities

The Secretary of Energy must study and compile existing science (including studies and data) to determine the risks or benefits presented by cumulative impacts of multiple offshore LNG facilities reasonably assumed to be constructed in an area of the Gulf of Mexico using the open-rack vaporization system. The study must be performed in consultation with the National Oceanic Atmospheric Administration, the Commandant of the Coast Guard, affected recreational and commercial fishing industries, and affected energy and transportation stakeholders.

The Secretary of Energy must verify the accuracy of available science and develop a science-based evaluation of significant short-term and long-term cumulative impacts, both adverse and beneficial, of multiple offshore LNG facilities reasonably assumed to be constructed in an area of the Gulf of Mexico using or proposing the open-rack vaporization system on the fisheries and marine populations in the vicinity of the facility.[42]

[30] Energy and Water Saving Measures in Congressional Buildings

The Architect of the Capitol, as part of the process of updating the Master Plan Study for the Capitol complex, must study and evaluate the energy infrastructure of the Capitol complex to determine how to augment the infrastructure to become more energy efficient.[43]

[31] Study of Availability of Skilled Workers

The Secretary of Energy must enter into an arrangement with the NAS under which the NAS will study the short-term and long-term availability of skilled workers to meet the energy and mineral security requirements of the United States. The study must include an analysis of the need for and availability of workers for the oil, gas, and mineral

---

[41] 2005 Energy Policy Act § 1827.

[42] 2005 Energy Policy Act § 1828.

[43] 2005 Energy Policy Act § 1829.

industries, of the availability of skilled labor at both entry level and more senior levels, and of recommendations for future actions needed to meet future labor requirements. The Secretary of Energy must submit to Congress a report that describes the results of the study no later than two years after August 8, 2005.[44]

[32] Review of Energy Policy Act of 1992 Programs

The Secretary of Energy must study the effect that titles III, IV, and V of the Energy Policy Act of 1992[45] have had on the development of alternative fueled vehicle technology, the availability of that technology in the market, and the cost of alternative fueled vehicles. The study must be completed no later than February 8, 2006.[46]

Upon completion of the study under this section, the Secretary of Energy must submit to Congress a report that describes the results of the study and includes any recommendations of the Secretary of Energy for legislative or administrative changes concerning the alternative fueled vehicle requirements under titles III, IV and V of the Energy Policy Act of 1992.[47]

[33] Study on the Benefits of Economic Dispatch

The Secretary of Energy, in coordination and consultation with the states, must study the procedures currently used by electric utilities to perform economic dispatch.[48] The study must identify possible revisions to those procedures to improve the ability of nonutility generation resources to offer their output for sale for the purpose of inclusion in economic dispatch. The study also must evaluate the potential benefits to residential, commercial, and industrial electricity consumers nationally and in each state if economic dispatch procedures were revised to improve the ability of nonutility generation resources to offer their output for inclusion in economic dispatch.

[44] 2005 Energy Policy Act § 1830.

[45] 42 U.S.C. § 13211 *et seq.*

[46] 2005 Energy Policy Act § 1831.

[47] 2005 Energy Policy Act § 1831.

[48] For this purpose, the term "economic dispatch" means "the operation of generation facilities to produce energy at the lowest cost to reliably serve consumers, recognizing any operational limits of generation and transmission facilities." 2005 Energy Policy Act § 1832(b).

The Secretary of Energy is required to submit annual reports to Congress and the states on the results of this study, including recommendations to Congress and the states for any suggested legislative or regulatory changes. The first report must be submitted no later than November 8, 2005. Subsequent reports must follow on a yearly basis.[49]

[34] Renewable Energy on Federal Land

The Secretary of the Interior must enter into a contract with the NAS under which the NAS will study the potential of developing wind, solar, and ocean energy resources (including tidal, wave, and thermal energy) on federal land available for those uses under current law and the outer Continental Shelf, assess any federal law (including regulations) relating to the development of those resources that is in existence on August 8, 2005, and recommend statutory and regulatory mechanisms for developing those resources. The Secretary of the Interior must submit to Congress the results of this study no later than two years after enactment.[50]

[35] Increased Hydroelectric Generation at Existing Federal Facilities

The Secretary of the Interior, the Secretary of Energy, and the Secretary of the Army must jointly study the potential for increasing electric power production capability at federally owned or operated water regulation, storage, and conveyance facilities. The study must include identification and description in detail of each facility that is capable, with or without modification, of producing additional hydroelectric power, including estimation of the existing potential for the facility to generate hydroelectric power.

The Secretary of the Interior, the Secretary of Energy, and the Secretary of the Army must submit a report on the findings, conclusions, and recommendations of the study to the Committees on Energy and Commerce, Resources, and Transportation and Infrastructure of the House of Representatives and the Committee on Energy and Natural Resources of the Senate. The joint report is due no later than 18 months after August 8, 2005.[51]

---

49 2005 Energy Policy Act § 1832.

50 2005 Energy Policy Act § 1833.

51 2005 Energy Policy Act § 1834.

[36] Split-Estate Federal Oil and Gas Leasing and Development Practices

The Secretary of the Interior must undertake a review of the current policies and practices with respect to management of federal subsurface oil and gas development activities and their effects on the privately owned surface. The Secretary of the Interior's review must be made in consultation with affected private surface owners, oil and gas industry, and other interested parties.

The Secretary of the Interior must report the results of such review to Congress no later than February 8, 2006.[52]

[37] Resolution of Federal Resource Development Conflicts in the Powder River Basin

The Secretary of the Interior must review federal and state laws in existence on August 8, 2005, in order to resolve any conflict relating to the Powder River Basin in Wyoming and Montana between the development of federal coal and the development of federal and non-federal coalbed methane. The Secretary of the Interior is required to submit to Congress a report that describes methods of resolving such a conflict and identifies a method preferred by the Secretary of the Interior, including proposed legislative language, if any, required to implement the method. The report is due no later than February 8, 2006.[53]

[38] National Security Review of International Energy Requirements

The Secretary of Energy must study the growing energy requirements of the People's Republic of China and the implications of such growth on the political, strategic, economic, or national security interests of the United States. The Secretary must conduct the study in consultation with the Secretary of Defense and Secretary of Homeland Security. The study must include assessments of the type, nationality, and location of energy assets that have been sought for investment by entities located in the People's Republic of China and the extent to which investment in energy assets by entities located in

---

[52] 2005 Energy Policy Act § 1835.

[53] 2005 Energy Policy Act § 1836.

the People's Republic of China has been on market-based terms and free from subsidies from the People's Republic of China. Additionally, the study must assess the effect of investment in energy assets by entities located in the People's Republic of China on the control by the United States of dual-use and export-controlled technologies, including the effect on current and future access to foreign and domestic sources of rare earth elements used to produce such technologies and the relationship between the government of the People's Republic of China and energy-related businesses located in the People's Republic of China.

The Secretary of Energy also must examine United States energy policy and foreign policy as it relates to ensuring a competitive global energy market and the relationship between the United States and the People's Republic of China as it relates to pursuing energy interests in a manner that avoids conflicts. Finally, the study must include a comparison of the appropriate laws and regulations of other nations to determine whether a United States company would be permitted to purchase, acquire, merge, or otherwise establish a joint relationship with an entity whose primary place of business is in that other nation, including the laws and regulations of the People's Republic of China.

The Secretary of Energy, in consultation with the Secretary of Defense, is required to report to the President and the Congress on the findings of the study and any recommendations these Secretaries consider appropriate. The report is due no later than 120 days after August 8, 2005.

Section 1837(c) of the Act prohibits any instrumentality of the United States vested with authority to review a transaction that includes an investment in a United States domestic corporation from concluding a national security review related to an investment in the energy assets of a United States domestic corporation by an entity owned or controlled by the government of the People's Republic of China for 21 days after the report to the President and the Congress, and until the President certifies that he has received the report described above.[54]

---

[54] 2005 Energy Policy Act § 1837.

[39] Used Oil Re-Refining Study

The Secretary of Energy must undertake a study of the energy and environmental benefits of the re-refining of used lubricating oil. The Secretary of Energy must issue a report to Congress no later than November 8, 2005. The Secretary of Energy is required to include in its recommendations the specific steps that can be taken to improve collections of used lubricating oil and to increase re-refining and other beneficial re-use of such oil. The report and study must be prepared in consultation with the Administrator of the EPA.[55]

[40] Transmission System Monitoring

The Secretary of Energy and the FERC must study the steps to be taken to establish a system to make available to all transmission system owners and Regional Transmission Organizations[56] within the Eastern and Western Interconnections real-time information on the functional status of all transmission lines within such Interconnections. The FERC is required to assess technical means for implementing the transmission information system and to identify the steps the FERC or Congress must take to require the implementation of such system.

The Secretary of Energy and the FERC must submit a report of this study to Congress no later than February 8, 2006.[57]

[41] Report Identifying and Describing Status of Potential Hydropower Facilities

The Secretary of the Interior, acting through the Bureau of Reclamation, must submit to the Committee on Resources of the House of Representatives and the Committee on Energy and Natural Resources of the Senate a report identifying and describing the status of potential hydropower facilities included in water surface storage studies undertaken by the Secretary of Energy for projects that have not been completed or authorized for construction. The report is due no later than November 8, 2005.[58]

---

[55] 2005 Energy Policy Act § 1838.

[56] As defined in the Federal Power Act 16 U.S.C. § 791a *et seq.*

[57] As of November 23, 2005, the FERC had not initiated efforts to produce the study and required report. See the FERC website at www.ferc.gov/legal/maj-ord-reg/fed-sta/ene-pol-act.asp.

[58] 2005 Energy Policy Act § 1840.

INDEX

[References are to sections.]

A

B

[References are to sections.]

C

[References are to sections.]

[References are to sections.]

D

[References are to sections.]

[References are to sections.]

[References are to sections.]

[References are to sections.]

[References are to sections.]

F

[References are to sections.]

[References are to sections.]

[References are to sections.]

I

L

[References are to sections.]

M

[References are to sections.]

N

[References are to sections.]

[References are to sections.]

O

[References are to sections.]

[References are to sections.]

[References are to sections.]

P

[References are to sections.]

[References are to sections.]

R

[References are to sections.]

[References are to sections.]

T

[References are to sections.]

[References are to sections.]

[References are to sections.]

[References are to sections.]

[References are to sections.]